Nanoscale Science and Engineering Forum

Held at the 2024 AIChE Annual Meeting

San Diego, California, USA
27-31 October 2024

ISBN: 979-8-3313-1681-5

Printed from e-media with permission by:

Curran Associates, Inc.
57 Morehouse Lane
Red Hook, NY 12571

Some format issues inherent in the e-media version may also appear in this print version.

Printed with permission by Curran Associates, Inc. (2025)

For permission requests, please contact AIChE
at the address below.

AIChE
120 Wall Street, FL 23
New York, NY 10005-4020

Phone: (800) 242-4363
Fax: (203) 775-5177

www.aiche.org

Additional copies of this publication are available from:

Curran Associates, Inc.
57 Morehouse Lane
Red Hook, NY 12571 USA
Phone: 845-758-0400
Fax: 845-758-2633
Email: curran@proceedings.com
Web: www.proceedings.com

TABLE OF CONTENTS

2D AND 3D NANOMATERIALS

AREA PLENARY: BIONANOTECHNOLOGY (INVITED TALKS)

AREA PLENARY: CARBON NANOMATERIALS (INVITED TALKS)

BIONANOTECHNOLOGY FOR GENE AND DRUG DELIVERY

BIONANOTECHNOLOGY FOR SENSORS AND IMAGING

BIONANOTECHNOLOGY GRADUATE STUDENT AWARD SESSION

CARBON NANOMATERIALS I: DISPERSION, SURFACE STRUCTURE, AND BIOINTERACTIONS

CARBON NANOMATERIALS: GRADUATE STUDENT AWARD SESSION

MACHINE LEARNING FOR NANOMATERIALS FOR ENERGY APPLICATIONS

NANOMATERIALS FOR ENERGY CONVERSION

NANOMATERIALS FOR ENERGY STORAGE

NANOSCALE SCIENCE AND ENGINEERING FORUM DIVISION PLENARY

NANOSCALE SCIENCE ENGINEERING FORUM (NSEF)

NANOSCALE SCIENCE ENGINEERING FORUM (NSEF): FACULTY CANDIDATE

PLENARY SESSION FOR NANOMATERIALS FOR ENERGY APPLICATIONS (INVITED TALKS)

POSTER SESSION: NANOSCALE SCIENCE AND ENGINEERING FORUM

POSTER SESSION: NSEF GRADUATE STUDENT POSTER COMPETITION

BIONANOTECHNOLOGY FOR SENSORS AND IMAGING II

484a Mxene-Stabilized Emulsions, Foams, and Composites

Huaixuan Cao, Miladin Radovic, Jodie Lutkenhaus, Micah Green, Emily Pentzer

Structured polymer composites have gained increasing attention due to their superior property enhancement (e.g., thermal, electrical conductivity) compared to their homogeneous counterparts, by designing the internal filler structures in the polymer matrix. The fabrication and design of structured polymer composites are still challenging and the common methods (e.g., solution casting, melt blending) have limited control over the internal filler structures. Pickering emulsion templating, in contrast, is an attractive approach to creating structured composites due to their well-defined interfaces, manageable structure and dimensions, and ease of scale-up. Among the common Pickering particles, MXenes are of great interest as they have the ability to not only stabilize emulsions but also introduce functional properties into structures, such as high electrical conductivity, high EMI shielding, and rapid radio frequency (RF) heating.

In this work, we focus on the development of MXene Ti3C2 Pickering emulsions in diverse fluid-fluid systems (e.g., oil-water and oil-oil) and their use as templates for fabrication of functional structured MXene-polymer composites. Pickering emulsions drive nanosheets to the fluid-fluid interfaces and subsequent localized polymerization creates diverse structured polymer composites (e.g., capsules, armored particles, and porous monoliths). The ability to access both aqueous and non-aqueous emulsion systems largely expands the possible polymer compositions. The Ti3C2 nanosheets are organized in these composites instead of being randomly distributed throughout. For instance, polymerization of the emulsion interfaces gives polymer shells with nanosheets embedded, polymerization of the dispersed phase gives polymer particles armored with nanosheets, and polymerization of the continuous phase gives porous monoliths with polymer struct and nanosheets coated pores. The incorporation of MXenes imparts functional properties into their structures for additional applications. For example, the MXene armored particles can be used as feedstock to fabricate segregated films for efficient EMI shielding applications at low MXene loadings due to the templated network within the polymers. MXene-polymer capsules and porous monoliths show excellent RF heating performance due to the highly locally conductive regions in these structures. The research work in this dissertation provides a simple platform to produce diverse structured MXene-polymer composites with well-controlled filler distribution, versatile compositions, and functional properties for potential advanced applications.

484b Copper-Decorated Carbon Nanofiber Mixed Matrix Polymer Membranes for Enhanced Permselectivity

Saketh Merugu

This presentation will explore the effects of multidimensional nanomaterials, specifically copper nanoparticles and high aspect ratio electrospun carbon nanofibers (CNF), on the interfacial properties of mixed matrix membranes when incorporated into a PVDF matrix. Remarkably, a mere 1wt% loading of Cu+CNF in the PVDF matrix led to a salt rejection rate of over 99.8% and a significant 64% rise in water vapor flux. The resulting alterations in attributes such as pore size, porosity, thickness, hydrophobicity, chemical heterogeneity, crystallinity, and thermal stability were verified using scanning electron microscopy (SEM), machine learning image segmentation, and Scattering and Wide-Angle X-scattering (SAXS and WAXS) methods. Furthermore, an autoregression forecasting model was developed using water vapor flux and salt rejection rate datasets to predict the long-term stability and performance of the membrane. The research offers important insights into the commercial potential of these enhanced membranes for MD applications.

484c Thermally Driven Stable Expansion of Hexagonal Boron Nitride through Intercalated Acid Treatment

Michael Prati, Megan A. Creighton

Hexagonal boron nitride (h-BN) is a 3D ceramic material composed of layered 2D boron nitride nanosheets (BNNS) interconnected by van der Waals forces. While isolated BNNS offer unique mechanical properties and applications, h-BN demonstrates exceptional thermal conductivity and stability in its 3D form. This study aims to extend previous methods of h-BN intercalation to isolate a stable expanded h-BN structure. Expanded graphite structures, currently utilized for various applications due to their excellent thermal properties, serve as a reference. However, graphite's electrical conductivity poses limitations. This study explores expanded h-BN as an alternative with superior thermal properties and electrical insulation, rendering it suitable for diverse applications, including electronics.

By employing sulfuric and phosphoric acid, we successfully intercalated bulk h-BN of varying lateral sizes, confirmed through X-ray diffraction (XRD) spectra. While smaller particles exhibit better intercalation, larger flakes offer potential for greater expansion due to a larger interface. Subsequently, the material undergoes exposure to temperatures ranging from 500-900 °C to investigate the extent of thermal shock. A preheated tube furnace with rapid transfer capabilities facilitates this process, trapping acid vapors and promoting layer expansion. Scanning electron microscopy (SEM) reveals rough and separated edges of the h-BN flakes compared to the parent material. XRD spectra analysis confirms distinct changes in lattice spacing, indicative of layer expansion. X-ray photoelectron spectroscopy (XPS) characterizes the composition, particularly identifying potential decomposition products from the acids. Exploiting this expanded h-BN form, applications such as binder-less h-BN thin films become feasible.

484d Enhanced Concrete Performance Using Graphene Produced Via a Continuous, Controllable, and Scalable Process

Andrius Patapas, Alperen Baki, Alalea Kia, Camille Petit, Jason Stafford, Omar Matar

The construction sector accounts for 40% of energy-related carbon emissions [1]. Concrete is the most commonplace construction material within the built environment, with cement, its main component, accounting for up to 8% of greenhouse gas emissions. Graphene is a two-dimensional (2D) material exhibiting high electronic, thermal, and mechanical properties. This material can serve as a concrete additive to not only improve its mechanical and durability performance, but also minimize its environmental impact by reducing the amount of cementitious material needed, thereby lowering carbon emissions. However, adding graphene reduces workability of concrete, and limits its ability to be placed into its mould/formwork. A more uniform graphene dispersion within concrete, and therefore better workable mixes, can be achieved by optimising the lateral size and thickness of graphene particles.

In the present work [2] (see Figure 1), we have developed an in-house turbulent Taylor-Couette (TC) flow system that generates the necessary conditions (strain rates) for effective and highly-controllable graphite exfoliation. We demonstrate our technological process for versatile and tonnage-production capability of pristine Few Layer Graphene (FLG) of required specification in terms of material quality and physical characteristics. The specification of FLG is determined through various characterisation techniques, such as AFM, UV-Vis, and Raman spectroscopy. In parallel, we showcase our novel on-line monitoring system based on optical spectroscopy for real-time measurement and on-the-fly control of the concentration and average number of atomic layers for the 2D material produced. This Industry 4.0 compatible process enables control over the whole value chain ensuring traceability, minimal waste, on-demand materials by design, adherence to ISO standards, and batch-to-batch conformity.

To understand the effect of FLG with different specifications on the fresh and hardened properties of cementitious mixes, we have conducted the following quantitative assessments:

- Isothermal calorimetry to determine the hydration profiles
- Slump flow to understand the rheology
- Compressive and flexural strength tests to investigate the structural performance
- Mercury Intrusion Porosimetry to characterise the porosity
- Scanning electron microscopy to study the surface morphology
- X-ray diffraction to analyse the physical properties
- Thermogravimetric analysis to determine the evolution of the chemical structure

By varying the exfoliation process conditions, multi-liter batches containing concentrated FLG dispersions with different specifications are produced particularly in terms of FLG concentration, thickness, and lateral size distributions. Different loading concentrations of FLG with best-performing specifications are blended into mortar samples and the above mentioned chemical and mechanical tests are conducted at different curing ages in order to understand the effect of FLG concentrations on the properties of cementitious materials. Quantitative assessment of the cured samples revealed improvements in key concrete performance characteristics, including compressive and flexural strength. These are directly compared with conventional concrete without graphene to determine the structural performance benefits resulting from the graphene additives. With such structural performance improvements in concrete, we showcase the positive environmental impact it has on the reduction of carbon dioxide emissions and of the amounts of precursors used and waste produced, increasing the overall sustainability levels within the construction sector by using 2D materials.

Acknowledgements: This research has received funding from the European Union's Horizon 2020 research and innovation programme under the Marie Sklodowska-Curie Grant Agreement No. 707340, and support from the EPSRC CDT in Advanced Characterisation of Materials (2018 NPIF grant EP/S515085/1) and UKRI Impact Acceleration Account (EP/X52556X/1).

References

[1] "Global Status Report for Buildings and Construction 2019 – Analysis - IEA," Apr. 2024, [Online; accessed 5. Apr. 2024]. [Online]. Available: https://www.iea.org/reports/global-status-report-for-buildings-and-construction-2019

[2] J. Stafford, U. Nwachukwu, U. Farooq, W. S. Favero, Silvia AND, H.-H. Chen, A. L'Hermitte, C. Petit, and O. Matar, "Real-time monitoring and hydrodynamic scaling of shear exfoliated graphene," 2D Materials, vol. 8, no. 2, 2021.

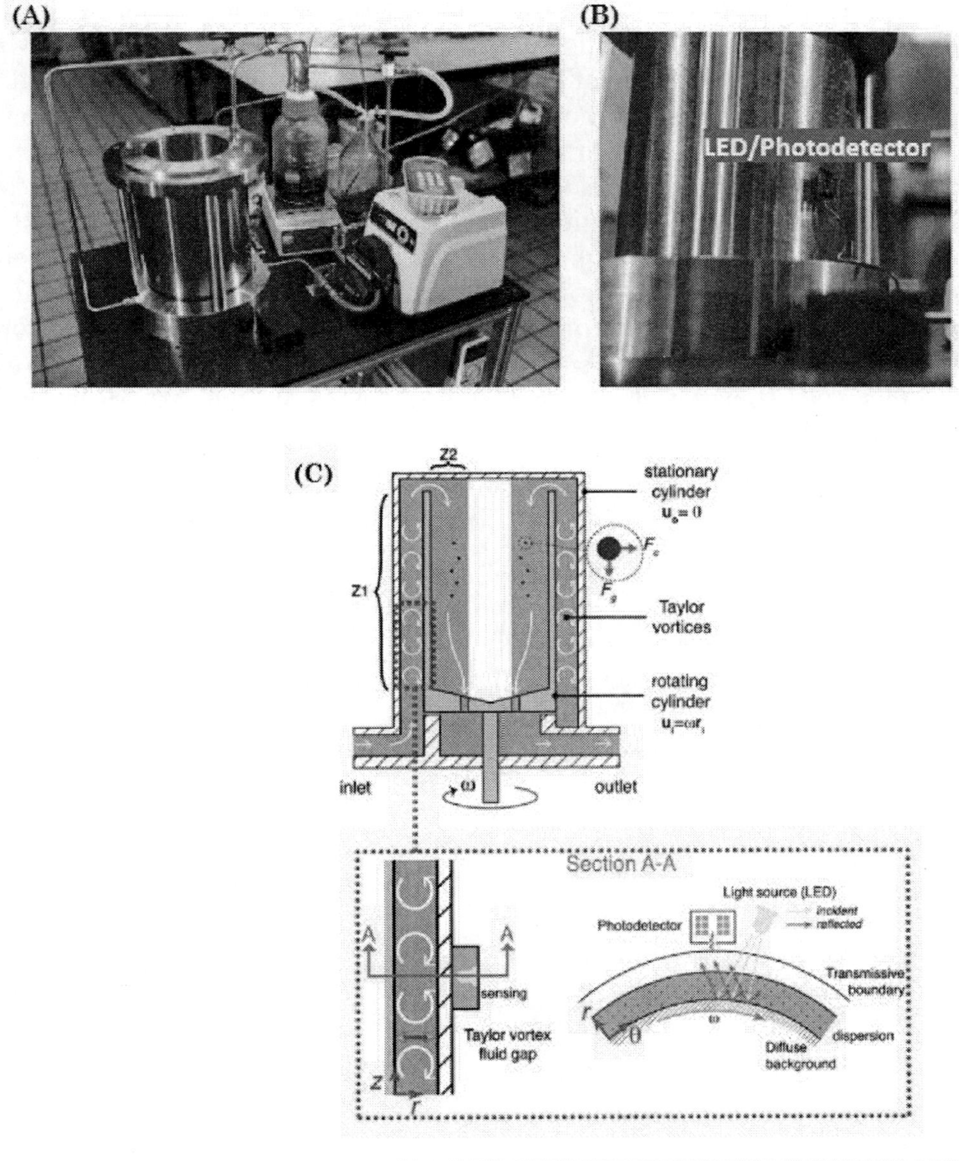

Figure 1: The manufacturing technology for graphite exfoliation: (a) Set-up photograph, (b) close-up photograph of the monitoring device (C) schematic of the TC and on-line monitoring devices.

484e Replacing Perfluoropolyether Media Lubricants in Hard-Disk Drives with Nanometer-Thick Ionic Liquid Coatings

Alan Tirado

Media lubricants are essential to the reliability and function of hard disk drives (HDDs). A good media lubricant has a low coefficient of friction (CoF), low surface tension for uniform surface coverage and self-healing, and can bond with the media surface to promote load-carrying capabilities. Additionally, data storage density can be increased if the lubricant has high thermal stability for heat-assisted magnetic recording (HAMR), and a low monolayer thickness to minimize head media spacing (HMS). However, the traditional HDD lubricants have been functionalized perfluoropolyethers (PFPEs) that lack the thermal stability required to fully enable HAMR technology, and the bulky molecular size of these polymers is a limiting factor in the reduction of the HMS. Ionic liquids (ILs) are promising candidates for this application due to their customizable nature that allows them to be designed with strong tribological properties, as well as their high thermal stabilities and relatively small size. Previous research in our lab has shown that while commercial ILs benefit from the aforementioned characteristics, they struggle to match the bonded thickness, low surface tension, and low CoF that the state-of-the-art PFPEs have achieved. Instead, we demonstrate the successful synthesis of a novel functionalized ionic liquid with highly fluorinated alkyl segments. In addition to the benefits of commercial ILs, our lubricant displays a comparable CoF to the PFPE Ztetraol, but has a lower surface tension and a shorter HMS. These results demonstrate that this novel IL has strong potential to be the next-generation lubricant that enables higher storage-capacity HDDs.

484f One-dimensional Contacts in Multilayer MoS2 Photodetectors

Zheng Yang

Since the early 21st century, the development of ultra-thin two-dimensional materials has introduced a new direction to human scientific understanding, thanks to their unique two-dimensional physical structures, exceptionally high surface-area-to-volume ratios, lightweight, high strength, and distinctive properties such as two-dimensional quantum confinement. In the field of photocatalysis, these materials have shown remarkable potential. Despite the substantial body of knowledge that has accumulated about 2D semiconductor photocatalysts to date, significant challenges remain in achieving widespread application of 2D semiconductor-based photocatalysis, including cost efficiency, degradation, charge transfer dynamics, and recyclability. Currently, most photocatalysts still require rare or expensive noble metals to enhance photocatalytic activity, significantly limiting their commercial-scale application. Moreover, the utilization of the exceptional surface-area-to-volume ratio of two-dimensional materials is often constrained by the use of dangling bonds at the edges of 2D materials or complex surface modifications, making it difficult to leverage the advantages of 2D materials over other substances.

Therefore, developing and creating novel structured two-dimensional material photocatalytic devices to achieve industrial-scale, low-cost, and high-efficiency conversion of solar energy poses a major challenge. In our research, we have successfully grown heterostructure thin films of single-layer two-dimensional material nanoribbons with alternating orientations through chemical vapor deposition, effectively leveraging the extremely high surface-area-to-volume ratio of two-dimensional materials for photocatalysis. Additionally, we achieved horizontally continuous alternating nanoribbons with each ribbon approximately 100 nm in width, greatly enhancing catalytic efficiency by maximizing the area of the depletion zone in the p-n junctions. The use of chemical vapor deposition for the practical fabrication of these materials could greatly facilitate the large-scale industrial production of 2D material-based photocatalytic materials.

484g Assessing the Quality of Large-Area Monolayer Graphene Grown on Liquid Copper for Size-Selective Ionic/Molecular Membrane Separations

Piran Kidambi

Monolayer graphene growth on liquid copper (Cu) has attracted attention due to advantages of a flat/smooth catalytic growth surface, high synthesis temperature (>1080 °C) as well as the possibility of forming graphene domains that are mobile on the liquid Cu with potential to minimize grain boundary defects and self-assemble into a continuous monolayer film. However, the quality of monolayer graphene grown on liquid copper and its suitability for size-selective ionic/molecular membrane separations has not been evaluated/studied. Here, we probe the quality of monolayer graphene grown on liquid Cu (via a metallurgical process, HSMG®) using Scanning Electron Microscope (SEM), High-resolution transmission electron microscope (HR-TEM), Raman spectroscopy and report on a facile approach to assess intrinsic sub-nanometer to nanometer-scale defects over centimeter-scale areas. We demonstrate high transfer yields of monolayer graphene (>93% coverage) from the growth substrate to polyimide track etched membrane (PITEM, pore diameter ~200 nm) supports to form centimeter-scale atomically thin membranes. Next, we use pressure-driven transport of ethanol to probe defects > 60 nm and diffusion-driven transport of analytes (KCl ~0.66 nm, L-Tryptophan ~0.7–0.9 nm, Vitamin B12 ~1–1.5 nm and Lysozyme ~3.8–4 nm) to probe nanoscale and sub-nanometer scale defects. Diffusive transport confirms the presence of intrinsic sub-nanometer to nanometer scale defects in monolayer graphene grown on liquid Cu are no less than that in high-quality graphene synthesized via chemical vapor deposition (CVD) on solid Cu. Our work not only benchmarks quality of graphene grown on liquid copper for membrane applications but also provides fundamental insights into the origin of intrinsic defects in large-area graphene synthesized via bottom-up processes for membrane applications.

Reference:

Romaniak G., Cheng P., Dybowski K., Kula P., Kidambi P.R., "Assessing the quality of large-area monolayer graphene grown on liquid copper for size-selective ionic/molecular membrane separations" Materials Research Express (2023) DOI: 10.1088/2053-1591/acefb2 part of themed collection: Emerging Investigators in Materials Science 2021

35a Manipulating Nanostructure Formation in Silk Fibroin-Based Materials to Stabilize Bioactive Cargo and Alter Material Properties

Whitney Stoppel

The quest to develop new, efficient ways to deliver bioactive cargo to meet the needs of the drug and protein delivery community never ends. Bioactive cargos, such as proteins or enzymes, can be challenging to deliver as each molecule comes with its own requirements, and thus the delivery and packaging strategy necessitates adjustments and optimization. Many material formats have been developed that can protect a bioactive cargo from destruction by the body until it reaches the desired destination. Substantial work in the liposome and polymersome communities highlights the complexity of the stabilization and delivery process, recognizing that this problem is multifaceted- stabilizing the cargo, stabilizing the formulation, targeting the delivery, and releasing the cargo at the right place and time.

In our work, we are interested in leveraging the nanostructures formed by the silk fibroin protein to stabilize bioactive cargos, creating another strategy to meet the needs of delivery systems. Silk fibroin, often isolated from *Bombyx mori* silkworms, is a protein with a high propensity to form secondary β-sheet structures. As engineers, we can leverage or control the formation of these nanoscale structures to kinetically trap bioactive cargos, such as extracellular matrix proteins, growth factors, and most recently, hemoglobin. For example, we have leveraged the phase separation of the silk fibroin protein under sonication conditions to entrap the hemoglobin, creating stable nano- and microparticles that can serve as hemoglobin based oxygen carriers. These particles range from 200 nm to >10 μm in diameter, as a function of silk fibroin concentration, the ratio of polymers within the system, and the molecular weight of silk fibroin.

Furthermore, we can also control the formation and organization of these structures to alter mechanical behavior of materials formed through modulation of the rate of β-sheet formation, ultimately impacting crystal size. The influence of these nanoscale protein domains on material properties allows for a wide range of material formats capable of entrapping bioactive cargo. This talk will highlight both the fundamentals of protein crystallization as it relates to material formulations and kinetic entrapment of bioactive cargos alongside some of the potential exciting biomedical applications for these materials.

35b Polymeric Nanotechnology Enables Versatile, Targeted Theranostics

Jessica Larsen

According to the World Health Organization, neurologic disorders account for the largest global burden through both early mortality and loss of independent life due to disability. Although there is a clear need for treatments that can enter the brain, there are a very limited number of medications available on the market. A major contributor to this fact is the presence of the blood-brain barrier (BBB), which prevents the passage of more than 98% of small molecule therapeutics from the blood into the brain tissue. The Larsen Lab works on polymeric biomaterials-based approaches to bypass and transport payloads through the BBB and the blood-nerve barrier. By understanding the pathophysiology of each disease, polymeric nanoparticles can be created to respond specifically to disease-based stimuli and promote natural healing processes in the brain. Polymersomes, self-assembled vesicles formed from amphiphilic block co-polymers, can enable simultaneous encapsulation and delivery of imaging agents and therapeutics, making them a valuable theranostic tool. In this talk, Professor Larsen will highlight some of her ongoing projects in this space, specifically focusing on the following: 1. Initial studies have developed enzyme-responsive polymersomes that can promote self-healing autophagic processes and enzyme restoration in neuropathic lysosomal storage disease model, GM1 gangliosidosis. Polymersomes made from hyaluronic acid (HA)-b-polylactic acid (PLA) are formed in a uniform size range using solvent injection formation techniques. Varying the HA molecular weight leads to control over release profiles of encapsulated enzyme beta-galactosidase. 2. These same enzyme-responsive polymersomes can co-deliver magnetic resonance imaging contrast agents with therapeutic proteins as an in situ neurologic tool. Furthermore, over time, degradation of HA-b-PLA polymersomes and release of the contrast agent leads to time-dependent loss of contrast. This demonstrates the potential for this tool to track drug release location. 3. Finally, pH-responsive polymersomes have been used as delivery vehicles for nerve-regenerative peptides isolated from the G3BP1 protein. These peptide-encapsulating polymersomes have been shown to promote neurite outgrowth in dorsal root ganglia cultures with neuronal specificity. Furthermore, the use of targeting ligands enables intramuscular injection in vivo in a rat model of sciatic nerve injuries. All of these projects demonstrate the versatility of the polymersome as a platform for important neural applications.

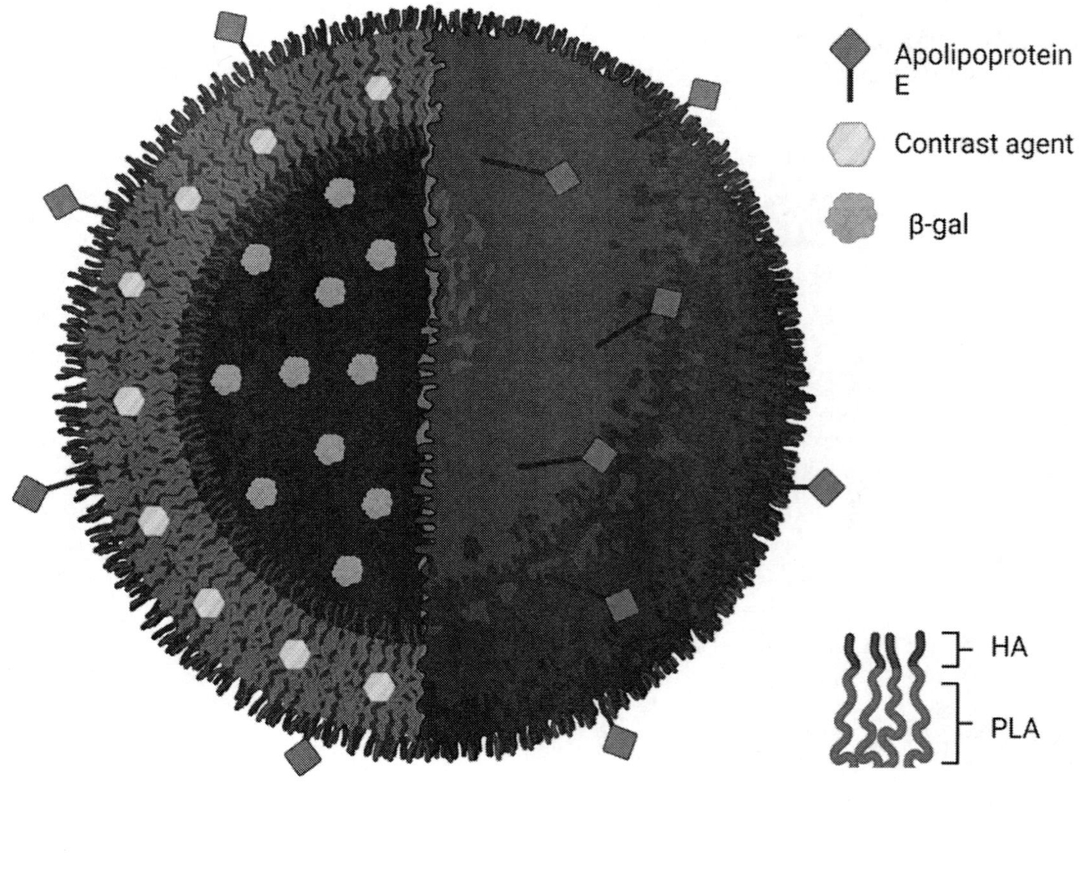

35c Nanoengineering Gone Viral: Plant Viruses Against Cancer

Nicole Steinmetz

Nanoscale engineering is revolutionizing the way we detect, prevent and treat diseases. Viruses are playing a special role in these developments because they can function as prefabricated nanoparticles. We utilize and build-upon the high-precision assemblies of plant viruses and repurpose them as platform technologies targeting health applications. We have developed a library of plant virus-based nanoparticles and through structure-function studies we are beginning to understand how to tailor these materials appropriately for immunotherapy applications. I will discuss the application and development of cowpea mosaic virus as a drug candidate for cancer immunotherapy; we demonstrate potent and tissue/tumor-type agnostic potency against metastatic tumors in mouse tumor models and canine cancer patients. Efficacy is demonstrated as solo-therapy as well as combination therapy or use in the neo/adjuvant setting.

241a Plant Protecting Plant: Investigating A Novel Biographene Growth Method On Fe For Corrosion Protection

Hema Ramsurn

Corrosion, a natural phenomenon, involving simultaneous oxidation and reduction reactions, leads to enormous economic losses worldwide. Several industrial applications including piping, cooling systems, boilers, heat exchangers, water treatment plants, etc. use metallic materials in the presence of saline media. Aggressive chloride ions in such media destabilize the passivation of even high chromium/nickel stainless steels as well as other normally-corrosion-resistant transition metals and lead to corrosion. Currently, metal surface protection can be achieved through several approaches including coating with other metals, chemical modification, anodizing, use of an oxide layer, and of organic layer/polymer coating. Recently, interest has piqued into using graphene as an ideal candidate for corrosion inhibition due to its unique properties like high electrical conductivity, transparency and chemical impermeability. Since its discovery, graphene has been successfully grown on Cu and Cu/Ni alloys using chemical vapor deposition (CVD). However, iron (Fe) is one of the most widely used metal in industry and chemical plants but to date, uniform graphene layers have not been able to be grown on this metal.

This work has the ambitious goal of making Fe corrosion-proof to increase its service life. The innovation lies in both the feedstock used and the carefully chosen process operating conditions. First, unlike the current state-of-the art method, Chemical Vapor Deposition (CVD), which mainly uses gaseous hydrocarbons, our process uses a solid renewable carbon source (biochar: carbon-rich residue left after hydrothermal carbonization (HTC) of biomass) to grow (bio)graphene on metal sheets like Fe. The mechanism is not deposition-based (CVD: weaker graphene-metal interaction) but rather dissolution-precipitation-based (stronger Fe-C affinity). This makes the adhesion of the graphene coating stronger, providing better corrosion protection when compared to the weaker CVD-grown graphene. Second, CVD is usually performed under high vacuum in the presence of H_2. Our process operates at atmospheric pressure and is H_2-free. Third, the Fe foils are used 'as is' without any pretreatment. These process conditions, being less harsh, are safer and more economical. Fourth, multilayer graphene, produced by our proposed process, has been shown to be better for corrosion prevention (CVD tends to produce single layer graphene). Fifth, any size of metal sheet can be "coated" with the graphene as this process is easily scaled up with the size of the furnace only and does not have to factor in the cost of vacuum and hydrogen to be used. Lastly, our process grows graphene on both sides of the foil simultaneously for better corrosion protection not only from the outside aggressive atmosphere but also from the internal corrosive material the pipe or vessel is carrying. We will be sharing some preliminary results obtained along with some analysis done using X-ray Diffraction (XRD), Raman spectroscopy and corrosion testing to explore our newly patented facile method to grow uniform (bio)graphene layers with high coverage on Fe for corrosion prevention.

241b There's Plenty of Room at the (Nano/Bio) Boundary

Ardemis Boghossian

The vast expansion of available synthetic biology tools has led to explosive developments in materials science. The increased accessibility of these tools has pushed the frontier of materials science into the field of engineering biological and even living materials. By coupling the tunable and robust optoelectronic properties of synthetic nanomaterials with the specificity and adaptability of biomaterials, one can re-purpose biology to fulfil needs that are otherwise intractable using traditional engineering approaches.

This presentation highlights applications in sensing and energy technologies that exploit the synergistic coupling of nanobio-hybrid materials at the boundary. This talk will discuss the development of bio-conjugated single-walled carbon nanotubes (SWCNTs) for near-infrared fluorescence sensing. We discuss recent advancements in applying synthetic biology approaches, such as directed evolution[1-3], xeno nucleic acid engineering[4-5], and protein mutagenesis[6], to control the optical properties of these synthetic nanoparticle sensors for a range of applications. This presentation will also discuss complementary efforts in re-purposing biological materials for electronic applications, like microbial fuel cells [7] and "living photovoltaics" [8-9], through concomitant genetic re-programming and nanomaterial engineering. These demonstrations exemplify but a few examples of nanomaterial bottlenecks that we can overcome through anti-disciplinary approaches.

References

- Lambert *et al.*, *Chem. Commun.* **55**, 3239–3242 (2019).
- P. Lambert *et al.*, *bioRxiv*, DOI: 10.1101/2023.06.13.544576 (2023).
- Rabbani *et al.*, *bioRxiv,* DOI: 10.1101/2023.09.07.556334 (2023).
- J. Gillen *et al.*, *J. Phys. Chem. Lett.* ***9***, 15, 4336–4343 (2018).
- J. Gillen *et al.*, *bioRxiv, DOI:* 10.1101/2021.02.20.428669 (2021).
- Zubkovs *et al.*, *Nanoscale Adv.* *4*, 2420-2427 (2022).
- Mouhib *et al.*, *Joule 7*, 2092–2106 (2023).
- Antonucci, *et al., Nat. Nanotechnol.* ***17***, 1111–1119 (2022).
- Antonucci, *et al., Photochem. Photobiol. Sci.* ***22***, 103–113 (2023).

Engineered Encapsulins As Advanced Drug Delivery Platforms and Catalytic Nanoreactors

Seokmu Kwon, Chemical Engineering, University of Michigan, Ann Arbor, MI

Abstract Text:
Protein nanocages have become important engineering tools for biotechnological and biomedical applications. Among them, encapsulins have recently received increased attention due to their favorable properties and selective protein packaging capabilities. Encapsulins are icosahedral protein compartments composed of a single type of shell protomer and are widely found in both bacteria and archaea. Encapsulins possess several key features that make them attractive engineering targets. First, encapsulins can specifically encapsulate dedicated cargo proteins. All native cargo proteins contain N-terminal domains or C-terminal peptides (TPs) necessary for efficient cargo loading during shell self-assembly (Figure 1). This feature – a dedicated and modular protein loading mechanism – has been widely utilized to package non-native cargo proteins into the encapsulin shell *in vivo* via simple genetic fusion of TPs to proteins of interest. Second, encapsulins exhibit diverse physicochemical characteristics useful for a variety of applications. For instance, encapsulins found in thermophilic organisms typically show extreme thermostability, while encapsulins encoded by acidophiles possess high acid tolerance. Lastly, ease of shell modification allows chemical conjugation, peptide insertion, and fusion of protein domains to the shell, resulting in potentially multifunctional protein nanocages.

Protein nanocage-based biomedical delivery applications have so far mainly focused on safely and efficiently transporting protein- or RNA-based therapeutics into target cells. However, the possibility of combining two types of specific therapeutic macromolecules – protein and RNA – in a single nanocage design has not been explored, which can have several advantages over delivering them separately. To seize this opportunity, we have engineered encapsulin shells to capture functional RNAs while retaining their inherent protein loading capacity. Positively charged peptides were genetically fused to the N-terminus of the encapsulin protomer, displaying the peptides at the luminal surface of the encapsulin shell (Figure 2). This modification enabled encapsulins to efficiently and size-selectively package RNAs *in vivo* through electrostatic interaction during shell self-assembly. Furthermore, the engineered encapsulin was capable of simultaneous loading of multiple functional RNAs and most importantly, it was capable of concurrent packaging of RNAs and specific cargo proteins *in vivo* (Figure 3). After demonstrating the RNA and protein co-packaging capabilities, we proceeded to further improve them by enabling the targeted encapsulation of select RNAs in a sequence-specific manner *in vivo*. Since the aforementioned system solely relies on electrostatic interactions for RNA encapsulation, various endogenous RNAs were also captured inside the encapsulin shell along with RNAs of interest. In order to make the system specific regarding RNA encapsulation, we fused the λN-peptide known to strongly bind specific RNA sequences (19 nt RNA hairpin sequence: BoxB RNA) to the N-terminus of the encapsulin protomer, replacing the positively charged peptide of the precursor system (Figure 4). In such a modified system, target RNAs of interest can be specifically captured inside the encapsulin shell by tagging RNA with the BoxB RNA sequence at the genetic level. Strikingly, Next Generation Sequencing-based analyses revealed that the λN-peptide-fused encapsulin almost exclusively encapsulates the co-expressed target RNA while excluding endogenous RNAs (98% of the total encapsulated RNA was the target RNA). Compared to other protein nanocage systems engineered for targeted RNA loading, the λN-peptide-fused encapsulin system showed superior specificity in terms of target RNA encapsulation. What makes this system even more valuable is the fact that such specificity is attained *in vivo*, making disassembling and reassembling of protein nanocages *in vitro* for target RNA loading unnecessary, which is an essential step for many other RNA encapsulation systems. On top of λN-peptide-fusion, in order to ensure efficient therapeutic cargo delivery upon endocytosis, we are in the process of additionally introducing a pH-responsive amphipathic peptide (GALA) along with several histidine point mutations into the encapsulin shell, aiming to trigger shell disassembly at low endosomal pH. The resulting highly specific RNA and protein co-encapsulation system with the added benefit of low pH-triggered disassembly, has the potential for simultaneous co-delivery of bioactive RNAs and proteins, improving therapeutic outcomes.

Encapsulins have also been exploited for the specific sequestration of various non-native enzymes with the goal of creating catalytically active nanoreactors with increased activity, stability, and longevity. However, the small native pores found in encapsulin shells often restrict the molecular flux of non-native enzyme substrates, cofactors, and products, decreasing the overall enzyme activity. Therefore, we engineered encapsulin shells with larger pores to allow more efficient access of non-native small molecules to the shell interior (Figure 5). In detail, we delete eight pore forming residues and added two glycine residues to allow flexibility. The resulting pore-mutated encapsulin (Mx_pmut) indeed improved the molecular flux across the shell, which was confirmed by SNAP-tag based fluorescence polarization (FP) assays. This improvement was characterized not only by faster molecular flux across the shell but also expansion of the upper size limit of substrate that can cross the shell, suggesting that a wider array of enzymes, particularly those utilizing large substrates or cofactors, can now be utilized with Mx_pmut based nanoreactors. This capability represents a significant advancement, as such enzymes could previously not be used with wild-type encapsulins due to substrate/cofactor inaccessibility. To demonstrate these improvements, we designed and tested two prototype nanoreactor systems based on a luciferase (NanoLuc) and a promiscuous alcohol dehydrogenase (AdhD), utilizing the relatively large substrate fluorofurimazine and redox co-substrate NADH, respectively (Figure 6). We showed that the turnover number for both enzymes significantly increased in Mx_pmut-based nanoreactors compared to wild type encapsulin (Mx_wt)-based nanoreactors. Lastly, we demonstrated the ability to co-encapsulate and produce functional multi-enzyme systems in Mx_pmut based nanoreactors. Specifically, we successfully implemented an efficient cofactor recycling system within Mx_pmut by co-encapsulating AdhD and the phosphite dehydrogenase PTDH (Figure 6). The improved performance consistently shown in nanoreactor systems based on Mx_pmut underscores the significance of pore size in determining overall nanoreactor performance, likely through controlling the molecular flux across the shell and thus the accessibility of substrates/cofactors and the release of products.

One of the potential concerns with increasing pore size is the loss of structural integrity and protective capacity of the shell. It has been shown that pores larger than >3 nm can allow small proteins like RNase to access the luminal space of the cage resulting in the loss of the protective capacity of the shell. Dynamic light scattering analyses, coupled with thermal ramp stability tests, showed that deleting pore-forming residues in Mx_pmut does not compromise the stability of the shell. Furthermore, encapsulated enzymes in Mx_pmut retained their activity upon proteolytic and bacterial cell lysate challenges, highlighting that the optimized porosity in Mx_pmut shells does not diminish their protective function (Figure 6).

In summary, we have engineered encapsulin nanocages to yield novel and versatile encapsulin-based drug delivery and high performance enzyme nanoreactor platforms with a broad future application range spanning biomedicine, biocatalysis, and bionanotechnology.

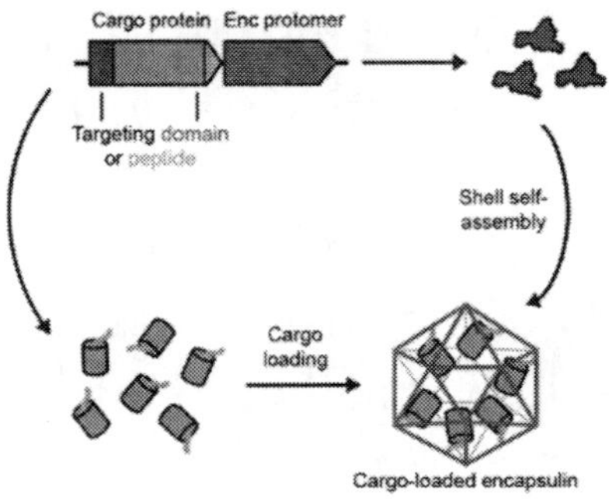

Figure 1. Schematic outlining encapsulin cargo loading and shell self-assembly.

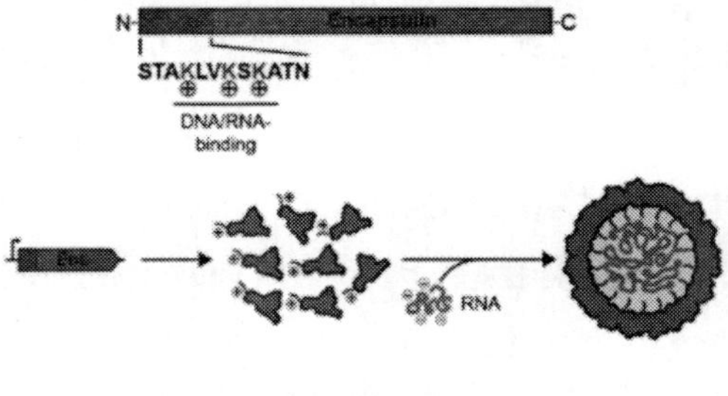

Figure 2. Schematic outlining the principle of *in vivo* RNA encapsulation by encapsulin fused to positively charged peptide.

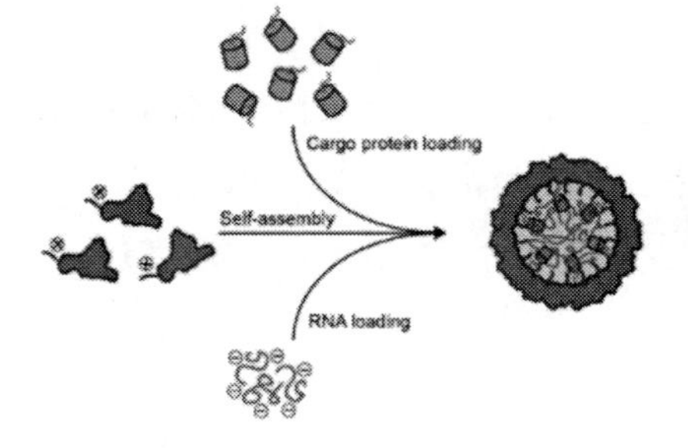

Figure 3. Schematic showing the concurrent packaging of RNAs and cargo proteins by encapsulin fused to positively charged peptide.

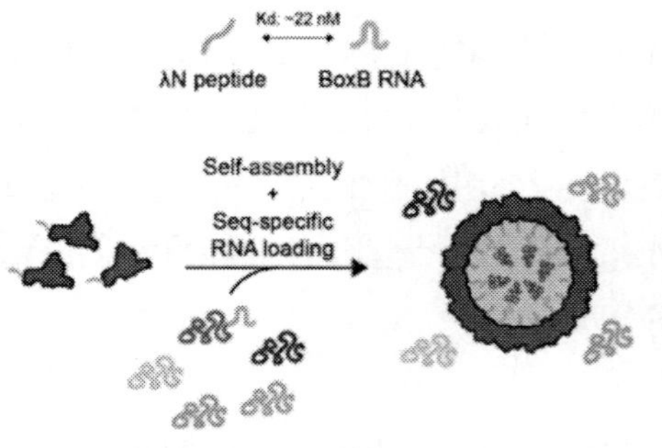

Figure 4. Schematic outlining the principle of sequence specific RNA encapsulation by encapsulin fused to λN peptide.

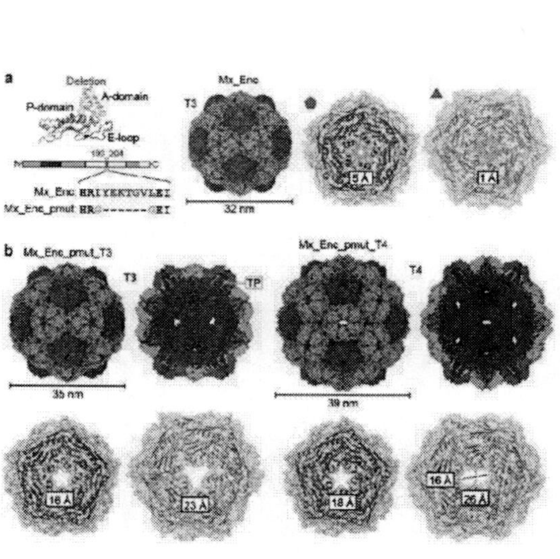

Figure 5. Structural characterization of our novel shell design. A) Design strategy for pore engineering (left) and encapsulin template used (right). B) Cryo-EM structures of novel encapsulin shells with increased pores size.

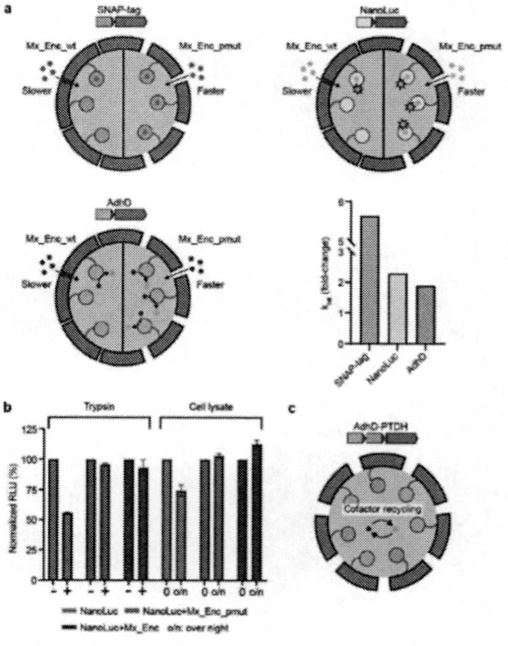

Figure 6. Functional characterization of our novel shell design. A) Prototyping of our novel nanoreactor design with three different enzymes. B) Stability comparison. C) Pore engineered shell with cofactor recycling system.

338b Self-Assembled Polymeric Oral Delivery Platform for Therapeutic Proteins

Fabiola Chapa Villarreal, Madi Schraad, Nikita Villegas, Sophia Castaneda, Nicholas A. Peppas

In recent years, monoclonal antibodies have become one of the most relevant therapies to treat autoimmune diseases. These conditions are characterized by the failure of an organism to tolerate its own cells and tissues which can lead to inflammation and tissue dysfunction [1]. However, because monoclonal antibodies have a low bioavailability and are highly susceptible to degradation, their administration is limited to parenteral routes [2]. Injections have a higher manufacturing cost and risk of infection, in addition to restricting their accessibility to patients [3]. Out of the numerous alternative administration routes that have been proposed, oral delivery offers the most advantages regarding costs, accessibility, and reduced discomfort, which increases patient compliance [2]. Nevertheless, the oral route has significant biological and physicochemical challenges such as the presence of proteolytic enzymes, drastic changes in pH, as well as the mucus and the epithelial cell layer [4]. All these challenges have to be overcome to successfully deliver monoclonal antibodies across the intestinal epithelium into the bloodstream. One approach to protect them from these barriers is to use polymeric nanocarriers based on block copolymers. Thus, in the present work we developed pH-responsive self-assembled polymersomes based on block copolymers to deliver monoclonal antibodies orally.

Diblock copolymers were synthesized using reversible addition−fragmentation chain-transfer polymerization and carbodiimide-mediated coupling reactions. Poly(methacrylic acid) (PMAA) and poly(ethylene glycol) (PEG) were selected due to their pH-responsive properties and stealth abilities, respectively. Different degrees of polymerization were examined to study the effect of the hydrophilic weight fraction on the micellar polymorphism. Analysis by Fourier-transform infrared spectroscopy and ^{1}H nuclear magnetic resonance spectroscopy confirmed the polymerization of PMAA and the successful addition of the PEG chain. Transmission electron microscopy confirmed the impact of the hydrophilic weight fraction on the micellar polymorphism as polymersomes and micelles were obtained. Since PMAA has a low pKa, strategies to improve the stability of the system at higher pH values have been implemented. The pKa of the system was increased by copolymerizing PMAA with hydrophobic comonomers, thus enhancing its suitability for oral delivery applications. In future studies, the loading and release capacity of the self-assembled systems will be studied as well as their cytocompatibility with Caco-2 cells. The potential to cross the mucus and the epithelial cell barrier will be analyzed using Caco-2 and HT-29/MTX co-culture. The development of these nanocarriers will help us achieve the oral delivery of monoclonal antibodies for the treatment of autoimmune diseases. By combining the material design of the nanocarriers with monoclonal antibodies to successfully deliver them through the oral route, the proposed work has the potential to impact the scientific community and the lives of many patients.

This work was supported by the NIH (R01-EB022025), the Cockrell Family Chair Foundation, the Office of the Dean of the Cockrell School of Engineering at the University of Texas at Austin (UT) for the Institute for Biomaterials, Drug Delivery, and Regenerative Medicine, and the UT-Portugal Collaborative Research Program. FAC-V acknowledges support from the CONACYT/ConTex Fellowship (Mexico).

References

[1] A. B. Shodeinde, A. C. Murphy, H. F. Oldenkamp, A. S. Potdar, C. M. Ludolph, and N. A. Peppas, "Recent Advances in Smart Biomaterials for the Detection and Treatment of Autoimmune

Diseases," *Advanced Functional Materials*, vol. 30, no. 37, p. 1909556, Sep. 2020, doi: https://doi.org/10.1002/adfm.201909556.
[2] Z. Antosova, M. Mackova, V. Kral, and T. Macek, "Therapeutic application of peptides and proteins: parenteral forever?," *Trends in biotechnology*, vol. 27, no. 11, pp. 628–635, Nov. 2009, doi: 10.1016/j.tibtech.2009.07.009.
[3] J.-P. Amorij *et al.*, "Towards tailored vaccine delivery: needs, challenges and perspectives," *Journal of Controlled Release*, vol. 161, no. 2, pp. 363–376, 2012.
[4] J. E. Vela Ramirez, L. A. Sharpe, and N. A. Peppas, "Current state and challenges in developing oral vaccines," *Advanced Drug Delivery Reviews*, vol. 114, pp. 116–131, 2017, doi: 10.1016/j.addr.2017.04.008.

338c Understanding the Role of Block-Length on Nanoparticle Assembly of Poly(2-hydroxyethyl methacrylate)-*Block*-Poly(lactic acid) *Via* Flash Nanoprecipitation

Erica Burnham, Rachel Pollard, Ashley Han, Lucy Stevens, Nathalie M. Pinkerton

Nanoparticle drug delivery systems (NDDS) predominantly utilize poly(ethylene glycol) (PEG) based stabilizing coronas due to its biocompatibility, stealth effect, and low opsonization. PEG has been increasingly reported to generate anti-PEG antibodies that contribute to an increased clearance of NDDS. This increased immunogenicity warrants evaluation of alternative stealth materials for NDDS. Poly(2-hydroxyethyl methacrylate) (pHEMA) is a hydrophilic, water-swellable, and FDA approved biocompatible polymer widely used in medicine, which can be utilized as an alternative material to generate the NDDS corona.

This work evaluates the use of pHEMA as a stabilizer for nanoparticles generated via the well-established Flash NanoPrecipitation (FNP) process. We synthesized a library of poly(lactic acid)-*block*-poly(2-hydroxyethyl methacrylate) (PLA-PHEMA) amphiphilic block copolymers (BCPs) with varying pHEMA molecular weight to elucidate the influence of molecular weight on nanoparticle assembly via FNP and to expand the polymeric stabilizer material repertoire for FNP. Assembly of polymeric nanoparticles with a pHEMA-corona has not yet been reported using FNP. Using a bifunctional initiator, ring opening polymerization (ROP) was used to initially synthesize a 10 kDa PLA block and subsequently atom transfer radical polymerization (ATRP) was used to synthesize pHEMA blocks with a range of molecular weights from 0.5 kDa to 5 kDa.

We observed a strong dependence of pHEMA molecular weight on nanoparticle assembly. BCPs with pHEMA blocks larger than 3 kDa were not able to form nanoparticles, while BCPs with smaller blocks robustly formed nanoparticles. This behavior can be explained by the dependence of pHEMA water solubility on the pHEMA molecular weight. The pHEMA-water chi parameter at molecular weights above 3 kDa approaches and then exceeds that of PLA-water. Thus pHEMA chains above 3 kDa decrease in water solubility and cannot sterically stabilize nanoparticles. With smaller pHEMA blocks, the nanoparticle size was tunable between 50 and 200 nm. Encapsulation of both vitamin E and rubrene aided in the assembly of nanoparticles with pHEMA molecular weights below 3 kDa. The cytotoxicity of 100 nm pHEMA stabilized nanoparticles was evaluated in F11 neuronal cells and found to have no significant cytotoxicity. Rubrene encapsulated pHEMA nanoparticles were found to be taken-up by these F11 cells.

We have demonstrated the use of pHEMA as a PEG-alternative stabilizer for FNP nanoparticles. The choice of pHEMA molecular weight is critical for both nanoparticle stability and assembly. This work ultimately expands the types of nanoparticle surfaces available for FNP and provides a framework for alternative biocompatible surfaces used in drug delivery nanoparticles.

338d Engineering MS2 Bacteriophage Virus-like Particle for Targeted Drug Delivery into Hepatocellular Carcinoma.

Daniel de Castro Assumpcao, Shiqi Liang, Neal Krishna Devaraj, Danielle Tullman-Ercek

Hepatocellular carcinoma, comprising 90% of liver cancers, exhibits a dismal five-year survival rate of 30%. Current treatment confronts limitations such as drug resistance and severe side effects. Nanotechnology, particularly Virus-Like Particles (VLPs), are a promising avenue for delivering therapeutics exclusively into the liver through targeted drug delivery, reducing side effects and enabling advanced treatments, such as gene therapy. To this end, we identified and engineered targeted drug delivery nanocarriers based on the native bacteriophage MS2 VLP. Initially, we screened sixteen MS2 VLP mutants that are assembly competent and contain an externally facing cysteine that could be conjugated with targeting ligands of interest via maleimide-thiol conjugation. Next, we selected variants that maintain the capsid quaternary structure after undergoing conjugation with maleimide-GalNAc, the ligand that binds to ASGPR receptors predominately encountered in hepatocytes. Finally, we demonstrated that MS2 VLP-GalNAc constructs are selectively internalized by cells that express ASGPR receptors. This research aims to advance the development of MS2 VLPs as a versatile platform for targeted drug delivery and the research outcomes are foundational for advances in therapeutic interventions for hepatocellular carcinoma.

338e Targeted siRNA Delivery to ACE2-Expressing Lung Cells Using MSC-Derived Extracellular Vesicles Functionalized with Receptor-Binding Domain of Sars-Cov-2 Spike Protein

Soha Alkhaldi, Ching-An Peng

Introduction: Extracellular vesicles (EVs) released from mesenchymal stem cells (MSCs) have been studied extensively for their potential therapeutic applications in lung injury. MSC-derived EVs could reduce lung inflammation by suppressing the production of pro-inflammatory cytokines. As a proof-of-concept study, MSC-derived EVs were tethered with the receptor-binding domain (RBD) of SARS-CoV-2 spike protein and carried small interfering RNA green fluorescent protein (siRNA GFP). The functionalized EVs are expected tospecifically target ACE2 receptors on A549 lung adenocarcinoma cells, thereby silencing GFP expression.

Methods: The gene encoding RBD of SARS-CoV-2 spike protein was amplified by PCR using pcDNA3-SARS-CoV-2-S-RBD-8his vector (Addgene), forward primer (5'-TAAGATATCGCCACCATGAAGCCATC-3', and reverse primer 5'-ACTAGAATTCAC-CGGATCCCTTTTTG G-3' (Integrated DNA Technologies). The purified PCR products treated with EcoRV and EcoRI restricted enzymes were inserted and ligated into pET30a (+)-streptavidin (SA) plasmid, and then expressed in BL21 (DE3) *E. coli*. The RBD-SA fusion protein purified by immobilized metal affinity chromatography was characterized by Western blot and bound to MSC-derived EVs biotinylated with Biotin-Cap-PE (Avanti Polar Lipids). The functionalized EVs, loaded with siRNA GFP usingCaCl-mediated precipitation and heat shock, were added to GFP-expressing A549 cells to evaluate the silencing efficacy at various times (8, 12, 24, and 48 hr).

Results: The purified RBD-SA fusion protein revealed MW of 37 kDa by Western blot analysis against His-tag and streptavidin monoclonal antibodies. The ACE2 receptors expressed on A549 cells were confirmed by detecting red fluorescent dye-conjugated anti-mouse IgG against anti-ACE2 primary antibody. For the ACE2-and GFP-expressing A549 lung cells targeted by RBD-tethered MSC-derived EVs encapsulated with siRNA GFP, the level of green fluorescence detected by plate reader and flow cytometry was notably dropped ove rtwo days. In fact, there was nominal no GFP expression 24 hr post-treatment. As a negative control, the EVs loaded with scramble siRNA illustrated high GFP expression over 48-hr cultivation.

Conclusions: The purified MSC-derived EVs tethered with RBD of SARS-CoV-2 spike protein and encapsulated with siRNA GFP could specifically target ACE2 receptors on GFP-expressing A549 lung cells and dramatically diminish the level of GFP expressed within the cells.

338f Injectable, Tunable Therapeutic Implant for Laryngeal Disease Treatment

Avery Zheng, TJ Waterkotte, Denzel Cruz, Tilahun Debele, Gregory Dion, Yoonjee Park

Laryngeal diseases, including vocal fold scarring and airway stenosis, require frequent local injections of corticosteroids. However, the effectiveness of current corticosteroid delivery methods is limited by factors such as short duration of action, discomfort, and costs associated with repeated procedures.

To address these limitations, we have developed two types of implants: (i) sustained-release capsule implants and (ii) light-activated capsule implants for local delivery of dexamethasone (Dex). For the sustained-release capsule implants, four different biodegradable polymers (polycaprolactone, poly(lactic acid), 90:10 poly(lactic-co-glycolide), and 50:50 poly(lactic-co-glycolide)) were used as capsule materials. The release profiles of Dex from these capsules were compared, and the 50:50 poly(lactic-co-glycolide) showed the fastest release. The permeability and partition coefficient of each polymer were measured, and it was found that the release of Dex through the capsule membrane was primarily due to "solution-diffusion" rather than diffusion through pores. The hydrophobicity of the polymer sheet was found to correlate with the drug release kinetics, with more hydrophobic polymers exhibiting slower release rates for the hydrophilic drug Dex. In vitro cell studies demonstrated the effective suppression of pro-inflammatory cytokine IL-6 levels using the Dex-loaded implant, indicating its potential therapeutic benefits with minimal toxicity.

Light-activated capsule implants were developed using light-activated liposomes encased within biodegradable polycaprolactone capsules. These liposomes, responsive to laser irradiation, facilitated controlled drug release, modulated by the duration of irradiation. Our experiments demonstrated successful drug release from these implants using a near-infrared (NIR) pulsed laser at 100 mW, connected via an optical fiber. When tested in a tissue phantom resembling laryngeal tissue in light transparency, the release kinetics exhibited a 4% reduction with every 2 mm increase in tissue depth. Additionally, in vivo studies on rabbit vocal folds revealed that the implants remained in place throughout the 14-day study period, with histology results indicating no adverse effects on the local tissue.

Overall, the development of the biodegradable polymer capsule for long-term release of Dex holds promise for improving the treatment of voice disorders, airway stenosis, and other related conditions.

338g Novel Injectable Thermoresponsive Hydrogels for Sustained Release of Antibody-Targeted DNA Nanocarriers As a Prophylactic Treatment for Secondary Cataracts

Camila Vardar, Mark Byrne

Cataracts are the second leading cause of blindness worldwide, with over 100 million cataract surgeries performed annually. Over 30% of adults and nearly all children develop secondary cataracts or clouding of the lens capsule following cataract surgery.Secondary cataracts, also known as posterior capsule opacification (PCO), originate from the differentiation of Myo/Nog cells into myofibroblasts, which contract and wrinkle the lens capsule. Current treatment strategies are insufficient and pose severe adverse effects; thus, a prophylactic treatment strategy delivered during cataract surgery is a considerable unmet need. Previous studies in a rabbit model suggest that a bolus dose of doxorubicin intercalated in DNA nanoconjugates containing the G8 monoclonal antibody (3DNA/Dox/G8), irrigated around the intraocular lens during cataract surgery resulted in the targeted depletion of myofibroblast precursors with minimal off-target effects. However, these studies reported only a short delay in the progression of PCO, instead of overall prevention.

To provide extended release of nanocarriers, injectable stimuli-responsive gels were designed using FDA-approved poly(lactic-co-glycolic acid)-b-poly(ethylene glycol)) (PLGA-PEG-PLGA) triblock copolymers and Poly-L-Lysine (PLL). PLL polycations provide multiple non-covalent complexation interactions between the homopolymer and 3DNA conjugates within the nanogel to control release. The addition of oppositely charged homopolymers forming co-assembled supramolecular hydrogels provided a novel platform for extended release exploiting non-covalent complexation interactions (US Patent pending, US20200038323A1). Incorporating cationic PLL chains of molecular weight in the 30-70kDa range into self-assembled PLGA-PEG-PLGA matrices controlled the release of negatively charged DNA-based nanocarriers for over 3 months. Our group has shown targeting specificity and cellular depletion sufficient to have therapeutic effects against PCO in vitro, in 28 days.

Our findings suggest that extended delivery of a low dose of 3DNA nanocarriers can have therapeutically relevant effects in the prevention of PCO. This system has the potential to impact every cataract surgery patient, decreasing the incidence of PCO and associated complications. Moreover, this gel formulation can be adapted to deliver various other types of therapeutics into the eye.

81b Enzyme Engineering to Produce SWCNT Nanosensors As a Generalizable Platform for Biochemical Imaging and Sensing

Shoichi Nishitani, Andrew Puglise, Deverin Smith, Markita Landry

Single-walled carbon nanotubes (SWCNT) are attractive candidates for near-infrared (nIR) fluorescent imaging and chemical sensing. However, to render SWCNT chemically-selective, rational design of the surface corona is required to achieve requisite sensitivity and selectivity to the target analyte. Recently, we reported on a facile sonication-based generation of glucose oxidase (GOx)-SWCNT conjugate for glucose detection [1]. Notably, we found that the analyte response mechanism does not involve catalytic oxidation of glucose, which motivated us to develop nanosensors using catalytically inactive GOx for enhanced reversibility and biocompatibility. Here, we study the generalizability of our approach in producing SWCNT-based nanosensors based on 1- facile direct probe tip sonication, and 2- with generically engineered enzymes, showing the broad applicability of this sonication-based approach as a generalizable strategy for generating rationally designed nanosensors capable of detecting and imaging various analytically-relevant biomolecules.

In this work, to test the generalizability of generating enzyme-SWCNT nanosensors, we synthesized conjugates by direct sonication of SWCNT with following enzymes: GOx, choline oxidase (ChOx), horseradish peroxidase (HRP), acetylcholineesterase, tyrosinase, cholesterol oxidase (CholOx), lactate oxidase, alcohol oxidase, xanthine oxidase, and galactose oxidase. Remarkably, six out of nine enzymes yielded stable suspensions of enzyme-SWCNT conjugates following probe-tip sonication, all of which successfully worked as nanosensors to detect their corresponding analytes. These nanosensors, in particular HRP-SWCNT and CholOx-SWCNT, exhibited excellent responses to their targets, with a maximum $\Delta F/F_0$ of up to 300 % and 100 %, respectively. H_2O_2 is known to quench SWCNT fluorescence, and H_2O_2 nanosensors previously developed with SWCNT exhibited a negative fluorescence modulation as the sensor output with some challenges towards nanosensor selectivity. We highlight that our approach enables HRP-SWCNT nanosensor generation, which provides a strong $\Delta F/F_0 = 300$ % and instantaneous turn-on response towards H_2O_2 within 2 s. Next, we sought to show that enzyme inactivation can produce SWCNT-based nanosensors that detect analytes without analyte consumption. To do so, we use ChOx as a model system to show that recombinant enzymes can be used as a generalizable approach to generating catalytically-inactive nanosensors. As expected, nanosensors prepared with mutant-ChOx exhibited responses comparable to those prepared with native ChOx, but without consuming the analyte or producing toxic byproducts such as H_2O_2. Thus, sonication-based physisorption of engineered enzymes to SWCNTs holds the potential to facilitate rapid nanosensor generation capable of detecting various biologically relevant molecules with reversibility and biocompatibility, thereby motivating their use in in vivo applications.

[1] Nishitani, S., Tran, T., Puglise, A., Yang, S. & Landry, M. P. Engineered Glucose Oxidase-Carbon Nanotube Conjugates for Tissue-Translatable Glucose Nanosensors. *Angew. Chem. Int. Ed Engl.* **63**, e202311476 (2024)

81c Tunable Clustering of Magnetic Nanoparticles Using Poly(amino acid) Corrals for Improved r_2 Relaxation

Geoffrey Bothun, Muzahidul Islam Anik, Adekunle Akinmola, Mohamed Kabil, Animesh Pan, Robert Mills, Carmen Scholz, Edward Walsh

Superparamagnetic iron oxide nanoparticle (SPION) clusters produce localized magnetic field inhomogeneities which could dramatically enhance T_2 relaxivity for magnetic resonance (MRI) and magnetic particle imaging (MPI). In this work, we report high and tunable T_2 relaxivity in hydrophobic SPION clusters self-assembled with amphiphilic polyethylene glycol-*b*-poly(*L*-Leucine) block copolymers, PEG-*b*-p(L-Leu)$_m$ where the PEG chains were terminated by methoxy (CH_3-O-PEG_{113}) or hydroxyl groups (HO-PEG_{77}). SPION clusters ranged in size up to 200 nm with the cluster diameter being dependent on the chain length of the hydrophobic p(L-Leu)$_m$ block. The r_2 relaxivity, determined from the SPIONs concentration dependence of T_2, of the assembled SPIONs increased with cluster diameter consistent with greater magnetic dipole-dipole interactions within larger clusters and the chemical exchange theory. Terminating the PEG segment with a hydroxyl group that has a high affinity for water resulted in SPION clusters formed with HO-PEG_{77}-*b*-p(L-Leu)$_m$ having a near 10-fold higher r_2 relaxivity compared to mPEG$_{113}$-*b*-p(L-Leu)$_m$ block copolymers, as well as commercially available SPION-based MRI contrast agents.

81d A POC Technology for Detecting HCV Directly from Untreated Plasma Using Rotational Brownian Motion of Immuno-Janus Particles

John Sinclair, Sonu Kumar, Satyajyoti Senapati, Hsueh-Chia Chang

Hepatitis C Virus (HCV) infections and associated liver diseases result in nearly 300,000 deaths annually. Recognizing this threat, the World Health Organization has set ambitious 2030 targets to reduce new HCV infections by 90% and related deaths by 65%. Despite the effectiveness of a 12-week antiviral drug regimen in treating HCV, access to effective diagnostic tools remains a major hurdle, particularly in developing countries where a significant portion of the HCV-related deaths take place. In these regions, diagnosis is predicated on an initial point-of-care (POC) antibody test of the patient's saliva or blood. A positive antibody result necessitates an analysis of viral load using reverse-transcription PCR. While PCR tests and mail-in services are common in developed nations, they are far less accessible in lower-income countries. Consequently, individuals are less likely to pursue follow-up testing, even if initial tests are positive for HCV antibodies. This limitation obstructs comprehensive diagnosis and subsequent treatment for many individuals.

Our laboratory addresses this challenge with the Immuno-Janus Particle (IJP) assay, designed for rapid depletion and quantification of viruses by immunocapture with antibodies on the IJP. The IJPs are micron-sized beads half-coated with a 30 nm layer of gold, undergo discernible Brownian motion with a "blinking" frequency inversely proportional to their diameter cubed. When viruses are captured by the bead, the effective diameter increases and reduces the rotation and blinking frequency. The platform is able to detect as few as 1000 viruses from a 10 μL of untreated plasma from a capillary after a skin prick, which approaches the sensitivity of PCR but without the necessary extraction protocol. The incubation time is less than 30 minutes because of the bulk binding. The IJP's size-based detection of eliminates the need for isolation, as soluble proteins are too small to create a signal and do not interfere with detection capabilities. An embedded microporous membrane allows the removal of blood cells by passive wetting flow in a lateral-flow chip. Moreover, its compatibility with standard smartphones enhances accessibility, as imaging the blinking beads requires only an inexpensive clip-on lens attached to a phone camera. By integrating our user-friendly MATLAB-based single-particle tracking code into a smartphone app, individuals can obtain the test results directly and transmit them wirelessly for region-wide epidemic control. These features hence render the platform a one-test definitive diagnostic test in the field that can be used in low-resource settings.

We are advancing the technology to a multiplex platform for detecting multiple blood-borne viruses in low-income countries, such as HIV, HBV and syphilis, with quantum dot labeling of the captured viruses. With labeling, magnetic IJPs will be used to enable an additional wash step to remove unbound quantum dots. A passive lateral flow design with a second wash flow will be tested.

81e Exploring the Corona Phase at Carbon Nanotube Surfaces - Corona Phase Molecular Recognition's Role in Nanosensor Development

Gabriel Sánchez-Velázquez, Michael S. Strano

The Corona Phase Molecule Recognition (CoPhMoRe) technique represents a significant advancement in molecular recognition, enabling the creation of selective interfaces around nanoparticles. This innovation has broadened the application range of single-walled carbon nanotube (SWNT) fluorescent sensors to include diverse analytes, such as hormones, reactive oxygen species, pharmaceuticals, and proteins. These nanosensors exhibit substantial efficacy in complex biological media, maintaining remarkable stability in various in vivo environments spanning both plant and animal systems. This work focuses on highlighting specialized experimental, theoretical, and computational tools essential for examining the corona phase, the molecular layer on the nanoparticle surface that is otherwise challenging to investigate. Experimentally, the Molecular Probe Adsorption (MPA) method is emphasized, which employs a fluorescent probe that becomes quenched upon interaction with the nanoparticle, thus enabling quantitative assessment of the solvent-exposed surface area through probe adsorption. Furthermore, in addressing the inverse problem of linking an analyte's structure to a synthetic corona on the SWNT surface, a theoretical framework capable of predicting the adsorbed configuration of a polymer on the nanotube surface is highlighted, along with other machine learning-based approaches. This work captures a comprehensive picture of the diverse sensing landscape accessible to CoPhMoRe, with an emphasis on the expanding body of work created to enhance the examination of the corona phase and enable the development of novel sensors for both in vivo and in vitro applications.

81f Electro-Spun Textiles with Encapsulated Single-Walled Carbon Nanotubes for Multiplexed Sensing and Hydrogel Based Therapeutics Delivery

Devleena Chowdhury, Amanda Pepler, Aceer Nadeem, Matthew Card, Ella Junge, Daniel Roxbury, Samantha Meenach

The process of wound healing induces the formation of new tissue which acts as an appropriate barrier from the external environment. For chronic and large-area wounds, it is desirable to monitor the healing process and simultaneously treat the wound, yet current strategies are limited.Therefore, creating alternate therapeutic and detection strategies for the timely monitoring and treatment of wounds is critical. Semiconducting SWCNTs with unique photophysical properties can act as fluorescent probes in the near-infrared (NIR) region. The photostability and long-term sensing abilities of single-walled carbon nanotubes (SWCNTs) with higher sensitivity in the tissue transparency window make them suitable to detect wound biomarkers. We have previously shown that SWCNTs functionalized with certain biopolymers can detect wound biomarker, e.g. H_2O_2,

and simultaneously impart colloidal stability and enhanced biocompatibility. We further encapsulated the nano sensors into an electro spun microfiber and demonstrated the insitu detection of H_2O_2. In addition to sensing, biodegradable polymers are considered ideal for therapeutics delivery applications as they encapsulate the therapeutics providing a sustained release. Polymers in the form of nanoparticles can improve encapsulation of therapeutics and provide sustained drug release profiles. Hydrogels have network structure which makes them smart drug delivery platforms. They can enable sustained release of therapeutics with slow degradation and controlled release. The release of the stimuli-responsive nanoparticles containing therapeutics facilitate the sustained release of therapeutics to the chronic wound over time. Here, we show our progress on a dual-use biocompatible micro fibrous platform with encapsulated SWCNT-based nano sensors and drug-loaded hydrogel for biomarker detection and wound therapy.

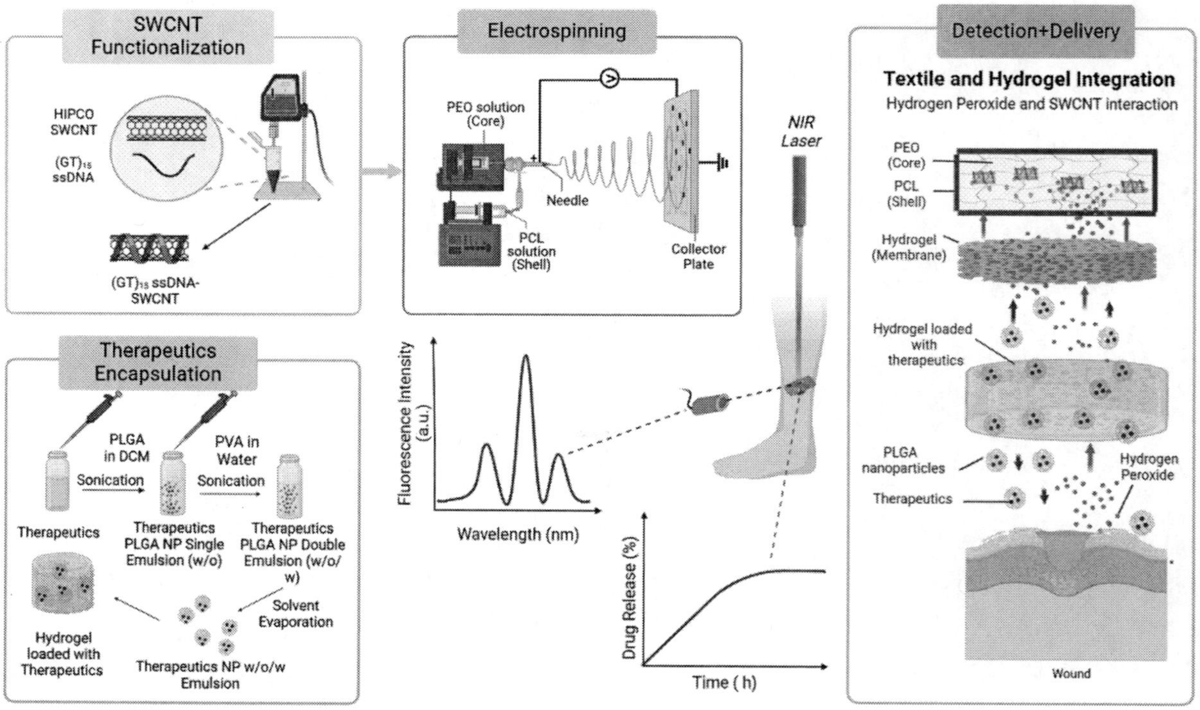

81g Characterization of a Nanocomposite Sweat Lactate Sensor for Continuous Monitoring on High-Level Athletes

Victoria Stege, Lisa Chambers, Madison Luthy, Michael Fulmer, Shelby Daniels, Chelsea Monty-Bromer

Presently, there is no established method for noninvasive and continuous monitoring of lactate levels during exercise. This is despite its significance in causing cramping, fatigue, and diminished performance in athletes due to anaerobic glucose metabolism. This study aimed to develop a fabric-based nanocomposite sensor capable of measuring sweat lactate levels during physical activity. The sensor fabrication involved treating non-woven nylon fabric with an aqueous dispersion of multi-walled carbon nanotubes (MWCNTs) functionalized with carboxylic acid groups. The sensor's active layer was synthesized using pyrrole and lactate oxidase. Characterization of MWCNT dispersion quality and particle size utilized UV-vis spectroscopy and Zetasizer analysis. Electrochemical properties were assessed through cyclic voltammetry of the sensor's response to synthetic sweat solutions containing lactate. Further testing involved analyzing sensor responses to spiked and synthetic sweat solutions containing common interferents like glucose, potassium, calcium, and sodium. Initial benchtop tests confirmed the sensor's ability to detect physiologically relevant levels of sweat lactate. Future evaluations will involve assessing the sensor's accuracy during on-body cycling trials.

276a Award Submission: Employing Novel Metasurface Nanopillars for the Detection of Antimicrobial Resistances Via Beta-Lactamase Activity

Jacob Waitkus, JaeWoo Park, Theodore Ndukaife, Ke Du, Sui Yang

Metasurfaces and metamaterial nanotechnology have undergone exponential growth in advancements which has opened the topic to a variety of mediums. This field has seen great interest in use as biosensors specifically, because of their high customizability, sensitivity, and often nondestructive sensing strategies.[1] Through precise control of the dimensions of subwavelength features, a variety of novel effects can be observed based on particle interactions. Due to alterations in the surrounding refractive index, or from plasmon resonance in the presence of photons, the unique surface properties of the nanoscale structures can undergo detectable changes.[2],[3] Fabrication of the novel silicon nanopillars described herein allowed for production of the intricate geometries capable of producing a desirable reflectance spectrum for the sensitive and specific detection of bound analytes, namely antibiotics and enzymes related to antimicrobial resistances (AMR). With the ever-expanding use of antibiotics in healthcare, combating bacterial resistance has emerged as a forefront issue.[4] Agar dilutions have become the established gold standard for the identification of antibiotic resistance among bacteria colonies.[5] This technique is extremely accurate and capable of identifying AMR in many bacteria samples simultaneously but is labor-intensive and expensive to perform with the equipment and trained personnel required. To overcome these issues, detecting the presence of β-lactamase enzymes allows for a simpler and cheaper strategy for observing antibiotic resistance to β-lactam class antibiotics in bacteria. Incorporating metasurface technologies for AMR detection will offer higher sensitivity than popular lateral flow assays, while maintaining the rapid result capabilities.

The 1 cm^2 samples used in this work were covered with the periodic nanopillar array, which consisted of 150 nm, 180 nm, or 200 nm diameter nanopillars at a customizable edge-to-edge distance varying from 155 nm to 240 nm. The height of the nanopillars were controlled at 110 nm or 210 nm by adjusting etching time and plasma composition, which allows for greater control over the pillar shape and the final spectra of the metasurface. This metasurface resulted from a facile fabrication process that only required DUV lithography with a negative photoresist/BARC layer, and a metal etch process with an accompanying strip step. The nanofeatures produced here pushed the capabilities of common DUV equipment, lending to specialized pillars that have been sparsely investigated. Through the combination of metallic nanoparticles with the nanopillars, simulations have suggested that a novel double-peak reflectance spectrum with a narrow FWHM below 10 nm can be observed, which is comparable to experimental findings for the silicon metasurface. These sharp peaks and valleys would increase the sensitivity of the overall system, resulting in simple and rapid detection of the antibiotic signatures.

The initial assay was developed for the sensing of target antibiotics via binding with the surface of the silicon nanopillars and surrounding planar surface. The silicon had been functionalized by a self-assembling monolayer consisting of an 11-aminoundecyltriethoxysilane (AUTES) monomer, which was confirmed by contact angle measurements. This molecule coated the silicon surface with a terminal amino group that preferentially bound to the carboxylic compound existing within the β-lactam antibiotic.[6] The antibiotic acted as a linker in this assay, allowing for specialized BSA-coated gold nanoparticles (BSANS) with a glutaraldehyde linker to be adhered to the functionalized silicon features. This glutaraldehyde molecule was essential as a spacer between the antibiotic and BSA, preventing the cephalexin antibiotic from being enveloped in the protein's binding domains. The coupling of the gold to silicon by the AUTES-antibiotic-glutaraldehyde chain allowed for plasmon resonant effects to be detected as 7 nm peak wavelength shifts in the double-peak reflectance spectra at picomolar concentrations of BSANS with nanomolar concentrations of cephalexin present. Alterations to the assay can be done for the detection of β-lactam-resistant bacteria through degradation of the antibiotic by the β-lactamase enzyme, preventing capture of nanoparticles on the nanopillar surface. By functionalizing the nanopillars with PBPs, which would only bind to intact β-lactam antibiotics, the presence of AMR can thereby be observed. The AUTES monomer in this assay would be replaced by a (3-Mercaptopropyl) triethoxysilane (MPTES) monomer which has a terminal thiol group, as opposed to an amino group, which is preferable for the capture of the PBP. Extensions of this work would lead to detection of these enzymes in a complex media such as milk and urine, where the BSA coating on the nanoparticles will aid in preventing non-specific protein absorption while identifying antibiotics and enzyme presence in a high noise environment. The metasurface technology developed here, in combination with the biological assay, would assist in further advancing AMR detection by simplifying the sensing strategy and speeding up result time compared to popular methods, while maintaining the high sensitivity and specificity required for clinical applications.

References:

[1] J. Hong, M. Su, and K. Zhao et al., Biosensors **13**, 327 (2023).

[2] T. Zhou, W. Ji, H. Fan et al., Biosensors **13**, 681 (2023).

[3] J. Hu, F. Safir, K. Chang et al., Nature Communications **14**, (2023).

[4] R. Uchil, G. Kohil et al., Journal of Clinical and Diagnostic Research **8**, 7 (2014)

[5] M. Weinstein, J. Patel, B. Limbago et al., Clinical and Laboratory Standards Institute **M07 11**, 38 (2018)

[6] Q. Li, T. Zhang, and L. Bian, Journal of Chromatography B **1014**, 90 (2016).

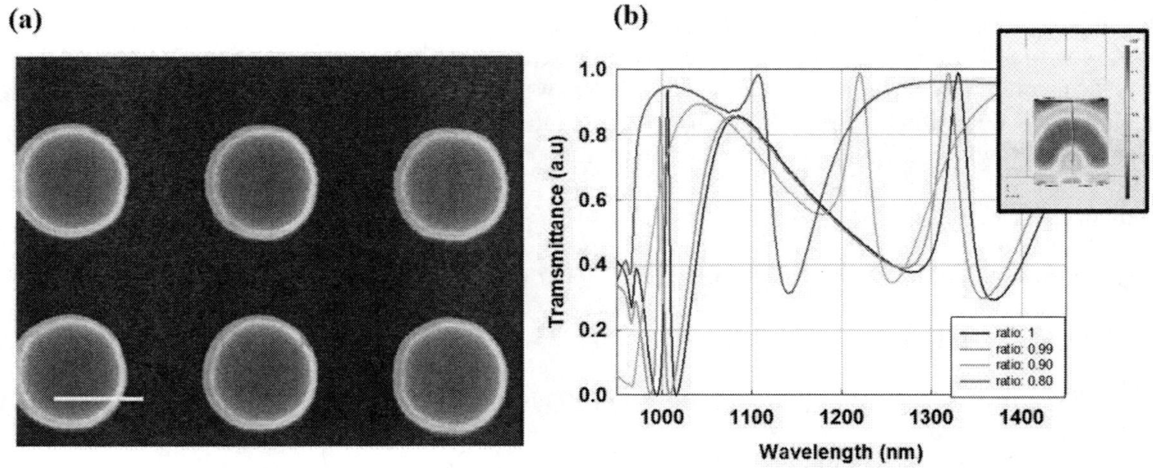

Figure 1: (a) SEM image of the nanopillar periodic array. (scale bar: 200nm) (b) Simulation of the double-peak reflectance spectrum for a changing ratio of the refractive index surrounding a nanopillar. Inset shows an electron heat map of a singular nanopillar coupled to adjacent pillar features.

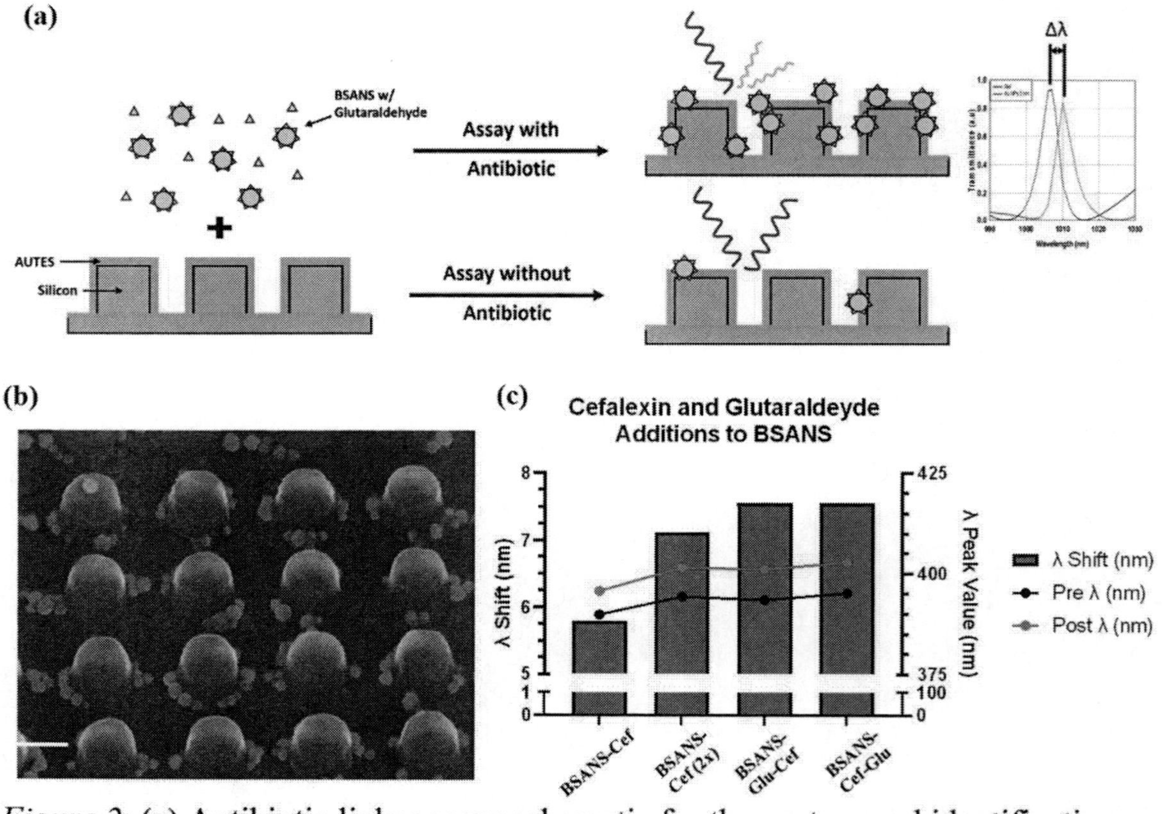

Figure 2: (a) Antibiotic linker assay schematic for the capture and identification of gold nanoparticles on the silicon metasurface. (b) SEM image for BSA-coated gold nanoparticles adhered to the AUTES layer on nanopillars. (scale bar: 200nm) (c) Bar plot showing the wavelength shift of the silicon nanopillars under different concentrations and order of cephalexin and glutaraldehyde.

Award Submission: Engineered Encapsulins As Catalytic Nanoreactors and Advanced Drug Delivery Platforms

Seokmu Kwon, Chemical Engineering, University of Michigan, Ann Arbor, MI

Abstract Text:
Protein nanocages have become important engineering tools for biotechnological and biomedical applications. Among them, encapsulins have recently received increased attention due to their favorable properties and selective protein packaging capabilities. Encapsulins are icosahedral protein compartments composed of a single type of shell protomer and are widely found in both bacteria and archaea. Encapsulins possess several key features that make them attractive engineering targets. First, encapsulins can specifically encapsulate dedicated cargo proteins. All native cargo proteins contain N-terminal domains or C-terminal peptides (TPs) necessary for efficient cargo loading during shell self-assembly (Figure 1). This feature – a dedicated and modular protein loading mechanism – has been widely utilized to package non-native cargo proteins into the encapsulin shell *in vivo* via simple genetic fusion of TPs to proteins of interest. Second, encapsulins exhibit diverse physicochemical characteristics useful for a variety of applications. For instance, encapsulins found in thermophilic organisms typically show extreme thermostability, while encapsulins encoded by acidophiles possess high acid tolerance. Lastly, ease of shell modification allows chemical conjugation, peptide insertion, and fusion of protein domains to the shell, resulting in potentially multifunctional protein nanocages.

Protein nanocage-based biomedical delivery applications have so far mainly focused on safely and efficiently transporting protein- or RNA-based therapeutics into target cells. However, the possibility of combining two types of specific therapeutic macromolecules – protein and RNA – in a single nanocage design has not been explored, which can have several advantages over delivering them separately. To seize this opportunity, we have engineered encapsulin shells to capture functional RNAs while retaining their inherent protein loading capacity. Positively charged peptides were genetically fused to the N-terminus of the encapsulin protomer, displaying the peptides at the luminal surface of the encapsulin shell (Figure 2). This modification enabled encapsulins to efficiently and size-selectively package RNAs *in vivo* through electrostatic interaction during shell self-assembly. Furthermore, the engineered encapsulin was capable of simultaneous loading of multiple functional RNAs and most importantly, it was capable of concurrent packaging of RNAs and specific cargo proteins *in vivo* (Figure 3). After demonstrating the RNA and protein co-packaging capabilities, we proceeded to further improve them by enabling the targeted encapsulation of select RNAs in a sequence-specific manner *in vivo*. Since the aforementioned system solely relies on electrostatic interactions for RNA encapsulation, various endogenous RNAs were also captured inside the encapsulin shell along with RNAs of interest. In order to make the system specific regarding RNA encapsulation, we fused the λN-peptide known to strongly bind specific RNA sequences (19 nt RNA hairpin sequence: BoxB RNA) to the N-terminus of the encapsulin protomer, replacing the positively charged peptide of the precursor system (Figure 4). In such a modified system, target RNAs of interest can be specifically captured inside the encapsulin shell by tagging RNA with the BoxB RNA sequence at the genetic level. Strikingly, Next Generation Sequencing-based analyses revealed that the λN-peptide-fused encapsulin almost exclusively encapsulates the co-expressed target RNA while excluding endogenous RNAs (98% of the total encapsulated RNA was the target RNA). Compared to other protein nanocage systems engineered for targeted RNA loading, the λN-peptide-fused encapsulin system showed superior specificity in terms of target RNA encapsulation. What makes this system even more valuable is the fact that such specificity is attained *in vivo*, making disassembling and reassembling of protein nanocages *in vitro* for target RNA loading unnecessary, which is an essential step for many other RNA encapsulation systems. On top of λN-peptide-fusion, in order to ensure efficient therapeutic cargo delivery upon endocytosis, we are in the process of additionally introducing a pH-responsive amphipathic peptide (GALA) along with several histidine point mutations into the encapsulin shell, aiming to trigger shell disassembly at low endosomal pH. The resulting highly specific RNA and protein co-encapsulation system with the added benefit of low pH-triggered disassembly, has the potential for simultaneous co-delivery of bioactive RNAs and proteins, improving therapeutic outcomes.

Encapsulins have also been exploited for the specific sequestration of various non-native enzymes with the goal of creating catalytically active nanoreactors with increased activity, stability, and longevity. However, the small native pores found in encapsulin shells often restrict the molecular flux of non-native enzyme substrates, cofactors, and products, decreasing the overall enzyme activity. Therefore, we engineered encapsulin shells with larger pores to allow more efficient access of non-native small molecules to the shell interior (Figure 5). In detail, we delete eight pore forming residues and added two glycine residues to allow flexibility. The resulting pore-mutated encapsulin (Mx_pmut) indeed improved the molecular flux across the shell, which was confirmed by SNAP-tag based fluorescence polarization (FP) assays. This improvement was characterized not only by faster molecular flux across the shell but also expansion of the upper size limit of substrate that can cross the shell, suggesting that a wider array of enzymes, particularly those utilizing large substrates or cofactors, can now be utilized with Mx_pmut based nanoreactors. This capability represents a significant advancement, as such enzymes could previously not be used with wild-type encapsulins due to substrate/cofactor inaccessibility. To demonstrate these improvements, we designed and tested two prototype nanoreactor systems based on a luciferase (NanoLuc) and a promiscuous alcohol dehydrogenase (AdhD), utilizing the relatively large substrate fluorofurimazine and redox co-substrate NADH, respectively (Figure 6). We showed that the turnover number for both enzymes significantly increased in Mx_pmut-based nanoreactors compared to wild type encapsulin (Mx_wt)-based nanoreactors. Lastly, we demonstrated the ability to co-encapsulate and produce functional multi-enzyme systems in Mx_pmut based nanoreactors. Specifically, we successfully implemented an efficient cofactor recycling system within Mx_pmut by co-encapsulating AdhD and the phosphite dehydrogenase PTDH (Figure 6). The improved performance consistently shown in nanoreactor systems based on Mx_pmut underscores the significance of pore size in determining overall nanoreactor performance, likely through controlling the molecular flux across the shell and thus the accessibility of substrates/cofactors and the release of products.

One of the potential concerns with increasing pore size is the loss of structural integrity and protective capacity of the shell. It has been shown that pores larger than >3 nm can allow small proteins like RNase to access the luminal space of the cage resulting in the loss of the protective capacity of the shell. Dynamic light scattering analyses, coupled with thermal ramp stability tests, showed that deleting pore-forming residues in Mx_pmut does not compromise the stability of the shell. Furthermore, encapsulated enzymes in Mx_pmut retained their activity upon proteolytic and bacterial cell lysate challenges, highlighting that the optimized porosity in Mx_pmut shells does not diminish their protective function (Figure 6).

In summary, we have engineered encapsulin nanocages to yield novel and versatile encapsulin-based drug delivery and high performance enzyme nanoreactor platforms with a broad future application range spanning biomedicine, biocatalysis, and bionanotechnology.

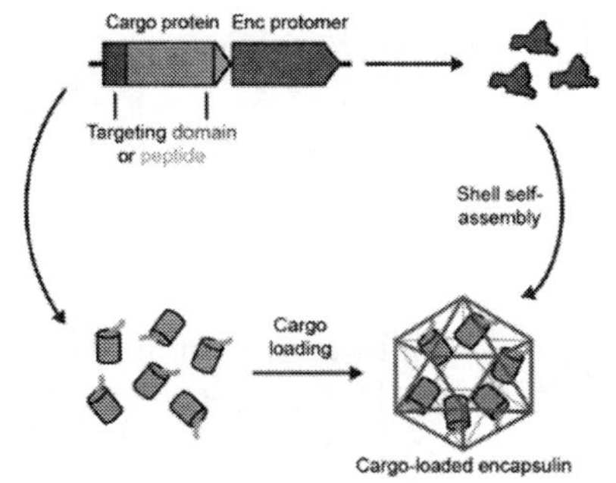

Figure 1. Schematic outlining encapsulin cargo loading and shell self-assembly.

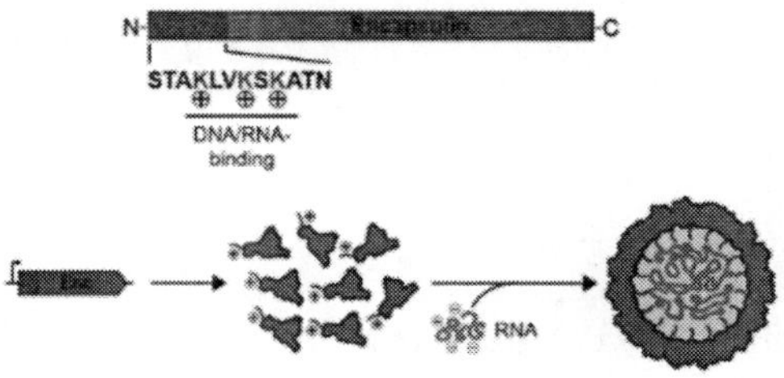

Figure 2. Schematic outlining the principle of *in vivo* RNA encapsulation by encapsulin fused to positively charged peptide.

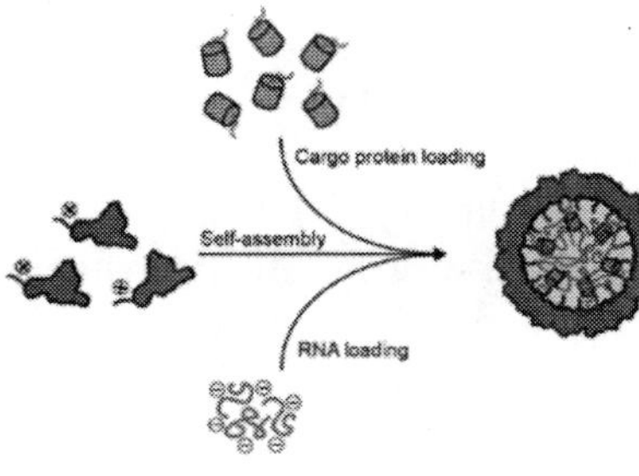

Figure 3. Schematic showing the concurrent packaging of RNAs and cargo proteins by encapsulin fused to positively charged peptide.

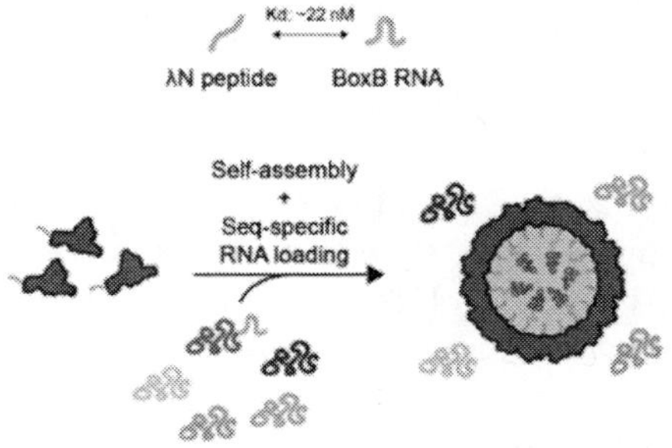

Figure 4. Schematic outlining the principle of sequence specific RNA encapsulation by encapsulin fused to λN peptide.

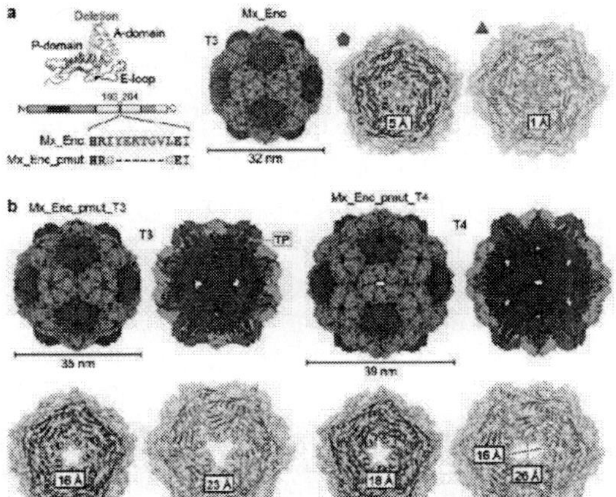

Figure 5. Structural characterization of our novel shell design. A) Design strategy for pore engineering (left) and encapsulin template used (right). B) Cryo-EM structures of novel encapsulin shells with increased pores size.

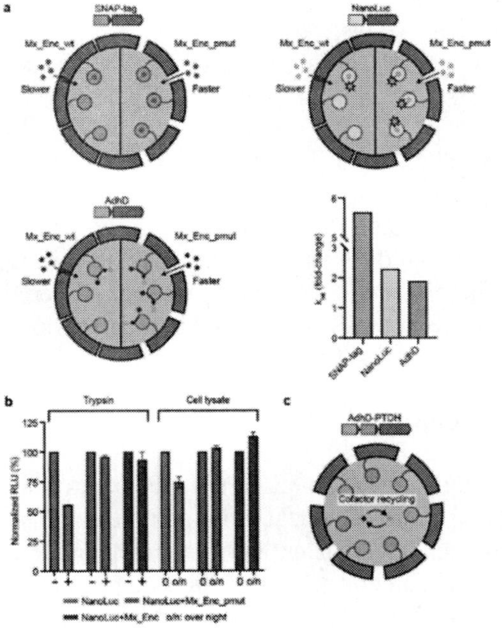

Figure 6. Functional characterization of our novel shell design. A) Prototyping of our novel nanoreactor design with three different enzymes. B) Stability comparison. C) Pore engineered shell with cofactor recycling system.

276c Award Submission: Impact of Surface Chemistry on Lipid Nanoparticle Transfection and Targeted Delivery

Namita Nabar, Tamara Dacoba, Gil Covarrubias, Denisse Romero-Cruz, Deeksha Kumaresh, Paula Hammond

Introduction

Lipid nanoparticles (LNPs) are potent delivery vehicles for multiple nucleic acid cargos, including messenger RNA (mRNA) and plasmid DNA. These systems have been used clinically for liver-targeted delivery of nucleic acid therapeutics, and their high transfection efficacy makes them valuable for treating diseases beyond the liver, such as sickle cell disease and extrahepatic cancers. However, targeted intravenous LNP delivery to extrahepatic spaces is challenged by multiple physiological parameters. NP surfaces adsorb negatively charged serum proteins, which induce aggregation and activate rapid sequestration by immune cells [1]. In particular, LNPs adsorb apolipoprotein E (ApoE), which enhances accumulation in the liver by binding to hepatocytic low-density lipoprotein (LDL) receptors [2]. Moreover, efficient gene delivery must mitigate off-target LNP uptake by more populous cell types. LNP surfaces for targeted extrahepatic gene delivery can therefore benefit from surface modifications.

The objective of this work is to investigate the transfection efficacy of a library of LNPs at varying lipid molar compositions, layered with surface polyanions using layer-by-layer electrostatic deposition [3]. Negative surface polyelectrolytes can not only minimize off-target uptake, but also facilitate covalent attachment of moieties to improve endocytosis and intracellular trafficking [4]. Rationally chosen polyelectrolyte chemistries can also impact nanoparticle biodistribution when delivered intravenously, which is advantageous for extrahepatic delivery.

Circulating LNPs are currently stabilized in physiological conditions by the presence of surface polyethylene-glycol (PEG) conjugated lipids. However, PEGylated nanocarriers may induce immune responses and reduce the efficacy of subsequent nanocarrier dosages [5]. Therefore, electrostatically adsorbed polyelectrolytes that confer stability advantages while minimizing immunogenicity may provide a viable alternative to PEGylation.

Methods

We encapsulated either GFP-expressing mRNA or plasmid DNA in LNPs, then deposited carboxylated surface polyanions composed of either saccharides (hyaluronic acid, HA), polypeptides (poly-L-aspartate, PLD; poly-L-glutamate, PLE) or acrylic groups (polyacrylic acid, PAA). Transfection efficacy of unlayered and layered LNPs was assayed *in vitro* in EL4 lymphoma cells, RAW 264.7 macrophages, and HEK293T epithelial cells via flow cytometry. Layered LNP intracellular trafficking patterns were assayed via confocal microscopy studies. LNP biodistribution, transfection, and circulation times were further assessed *in vivo* in naïve C57BL/6 mice.

Results and Discussion

Relative to unlayered cationic LNPs, LNPs layered with polyanions exhibited diameter increases of 5 to 10 nanometers and complete surface charge reversal (Fig. 1A-D, F). Encapsulation efficiency of both mRNA and pDNA remained consistent between unlayered and layered LNPs (Fig. 1E). *In vitro* studies in a diverse range of immune and epithelial cell lines demonstrated correlations between lipid molar compositions and LNP transfection. Layered LNPs retained transfection capacity, transfecting at comparable or higher rates to unlayered LNPs. Furthermore, the outer polyanion of layered LNPs modulated uptake and transfection in a cell-dependent manner; HA, for example, upregulated transfection in RAW macrophages but downregulated transfection in EL4 lymphoma cells. Tested *in vivo*, PLE and HA layers further improved LNP hepatic and splenic transfection; HA-layered LNPs demonstrated splenic specificity (Fig. 2A-B). Surface deposition of polyanions further stabilized non-PEGylated LNPs in salt and serum conditions, while preserving transfection *in vivo*.

These findings suggest that LNP surface chemistry can be modulated to improve targeting and uptake, to aid in the transfection of extrahepatic tissues.

Acknowledgements: Funding was provided by the Bill and Melinda Gates Foundation and the National Science Foundation.

References

[1] Kumar et al. *Chemical Reviews* 121, 11527-11652 (2021).
[2] Samaridou et al. *Advanced Drug Delivery Reviews* 154-155, 37-63 (2020).
[3] Nabar et al. *Proceedings of the National Academy of Sciences* 121(11), (2024).
[4] Correa et al. *ACS Nano* 14(2), 2224-2237 (2020).
[5] Ju et al. *ACS Nano* 16(8), 11769-11780 (2022).

Fig. 1

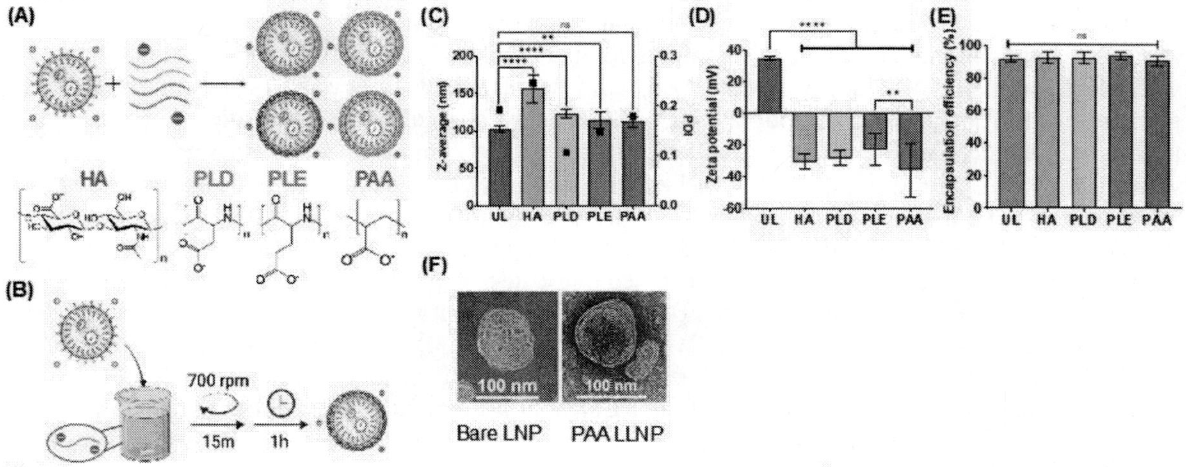

Fig. 2

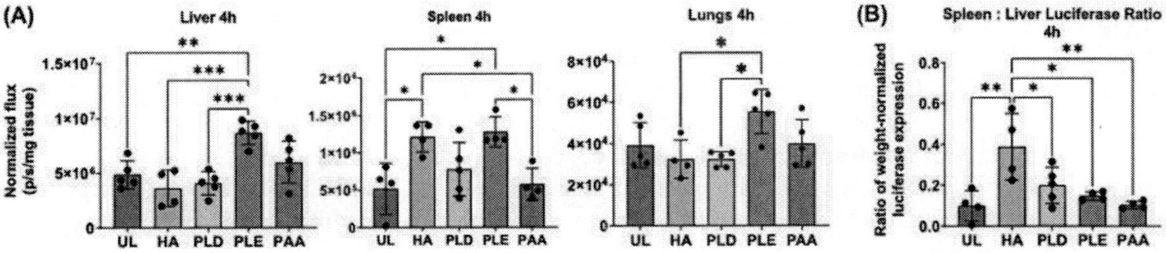

276d Protein-Loaded, Sustainably-Formulated Bacterial Cellulose Nanoparticles

Gabrielle Balistreri, Eleftheria Roumeli, Elizabeth Nance

Background: Many U.S. FDA approved materials in nanomedicine technologies involve chemically intensive syntheses and are difficult to manufacturer at the kilogram scale. Our prior work used copolymer poly(lactic-*co*-glycolic acid) (PLGA)-polyethylene glycol (PEG) to deliver drugs to the injured newborn brain. Although PLGA-PEG is a viable and effective platform, the synthetic material is limited by scalability and sustainability. Green engineering is one way to address the concerns of sustainability in formulation of nano-based delivery systems. Here, we develop bacterial cellulose nanoparticles (BCNPs) for sustainable drug delivery applications and motivated by the material's bio renewable sourcing and biocompatibility, which are important properties for nanomedicine applications. BCNPs are a therapeutically relevant formulation that has a reduced environmental impact, an overall eco-friendly life cycle (Figure 1) and can be implemented following greener engineering methodologies.

Methods: We applied surface functionalizations to BC to tune the surface functionality and charge density. To achieve the modifications, we utilize literature reported strategies, such as methylation, acetylation, and amination. We first used commercially manufactured After validating the functionalized alpha cellulose for positive and negatively charged cellulose matrices with Fourier transform infrared spectroscopy (FTIR) and X-ray crystallography (XRD), we next applied the same functionalization methods to BC pellicle. The BC pellicle was first collected from a kombucha co-culture media (bacteria and yeast) from the air and water interface, washed with sodium hydroxide and water for isolation and purifications steps, and then dissolved in small volumes of dimethylacetamide and lithium chloride. The BC pellicle was then functionalized using the same methods applied to alpha cellulose. After functional groups were confirmed, the BC pellicle was processed to create a small library of functionalized BCNPs for protein loading applications. The dissolved BC pellicle underwent a double emulsion formulation with polysorbate 80 surfactant solution to produce BCNPs. To make sure the functional groups were retained when making BCNPs from functionalized BC pellicle, we again performed FTIR and XRD. Slightly changing the pH during or post modification of the BCNPs allowed greater particle dispersion upon drying and redispersion in water. The particle size, zeta-potential, and polydispersity index (PDI) were measured by dynamic light scattering, NanoSizer, and electron microscopy.

Cytotoxicity of were evaluated in our established organotypic whole hemisphere (OWH) brain slice platform prepared from postnatal day 10 (P10) rat brains and compared to chitosan nanoparticles and alginate nanoparticles, field standards for polysaccharide-based nanoparticles. After plating 300 μm OWH slices on semi-permeable membranes in slice culture media, slices were cultured for 4 days in vitro (DIV). At 4 DIV, BCNPs were topically applied to slices at 3 doses: 5 μg, 25 μg, 75 μg. After 24 h of BCNPs exposure, slices were stained with propidium iodine (PI), a cell death stain, and TO-PRO-3, a positive nuclei stain. Slices were fixed in methanol and PI+ cells were imaged using confocal microscopy. A separate set of slices exposed to BCNPs were collected and RNA was isolated to perform qPCR analysis to measure cytokine response. Cellular activation markers that were assessed include: Ki67, CD68, GFAP, Synapsin, CD11b, Vim. Lastly, to demonstrate the ability to load catalase into BCNPs, we incorporated 3 drug loadings % of catalase at 5%, 10%, and 20% and measured the catalase activity by enzymatic assay to identify active drug in the BCNPs.

Results: We evaluated surface functionalization strategies in BC pellicle aiming to create functionalized nanoparticles, and benchmarked our methods against literature reported reactions where alpha cellulose was used as a validation for those reactions. Compared to alpha cellulose post-surface modifications, the BC pellicle was positively charged upon amination, while methylation, acetylation, and esterification increased the negative charge of BC, as represented in FTIR and XRD data. The surface chemistry consisted of reactions with the primary and secondary alcohol groups on the BC surface, where the surface chemistry dictated interfacial properties and interactions with other species. Double emulsion formulation allowed the development of BCNPs <100 nm in diameter, slightly negative zeta-potential, and have low PDI post surface modification of BC pellicle. These physicochemical properties align with our prior work for polymer nanoparticles that show promise for delivery of proteins to the brain. The surface modified BCNPs were compared to the chitosan, a positively charged polysaccharide, and alginate as controls for colloidal stability and protein interactions for potential drug loading applications. The biocompatibility results of BCNPs in ex vivo models show BCNPs are non-toxic at lower doses and do not initiate an inflammatory response in healthy brain slices. We also demonstrate that catalase, a low cost and readily available protein, can be incorporated onto surface modified BCNPs by utilizing positively charged or aminated BCNPs that facilitate electronic interactions with the available carbonyls on catalase. Other interactions that could occur between the BCNPs and catalase could potentially be hydrogen bonding stabilizing the platform based on the crystallinity and organization of the cellulose units observed by XRD.

Conclusion:

We formulated sub-100nm BCNPs with positive and negatively charged chemical modifications and benchmarked their protein loading and cytotoxicity in the brain against field standard alginate and chitosan nanoparticles. Our results establish a tunable and sustainably derived platform in nanomedicine for the application of protein-based delivery to the brain. The small library of functionalized BC can open opportunities for other utilizations of BCNPs and an eco-friendlier way to produce therapeutics using green engineering.

References:

1. Joseph, A., Nyambura, C. W., Bondurant, D., Corry, K., Beebout, D., Wood, T. R., Pfaendtner, J., Nance, E., *Formulation and Efficacy of Catalase-Loaded Nanoparticles for the Treatment of Neonatal Hypoxic-Ischemic Encephalopathy.* Pharmaceutics, 2021. **13**(8).
2. Balistreri, G.N., et al., *Bacterial cellulose nanoparticles as a sustainable drug delivery platform for protein-based therapeutics.* RSC Applied Polymers, 2024.
3. Tardy, B.L., et al., *Deconstruction and Reassembly of Renewable Polymers and Biocolloids into Next Generation Structured Materials.* Chem Rev, 2021. **121**(22): p. 14088-14188.
4. Nguyen, N.P., et al., *Brain Tissue-Derived Extracellular Vesicle Mediated Therapy in the Neonatal Ischemic Brain.* International Journal of Molecular Sciences, 2022. **23**(2): p. 620.
5. Foster, E.J., Moon, R. J., Agarwal, U. P., Bortner, M. J., Bras, Julien, C-E., Sandra., Chan, K. J., Clift, M. J. D., Cranston, E. D., Eichhorn, S. J., Fox, D. M., Hamad, W. Y., Heux, L., Jean, Bruno., Korey, M., Nieh, W., Ong, K. J., Reid, M. S., Renneckar, S., Roberts, R., Shatkin, J. A., Simonsen, J., Stinson-Bagby, K., Wanasekara, N., Youngblood, J., *Current characterization methods for cellulose nanomaterials.* Chemical Society Reviews, 2018. **47**(8): p. 2609-2679.

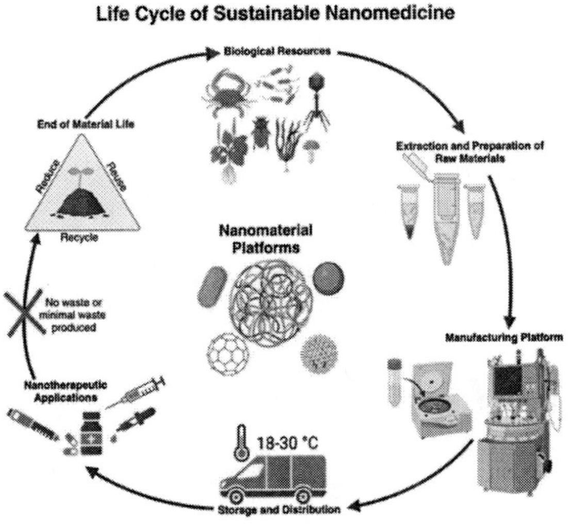

Figure 1. Environmentally friendly life cycle assessment for sustainable nanomedicine.

276e Award Submission: Designing a Magnetically Responsive Microbial Nanoculture System

Huda Usman, Tagbo Niepa, PhD

Background/Motivation: The study introduces magnetically functionalized nanocultures (MNCs) as a novel platform for the study of environmental microorganisms. The nanocultures are nanoliter-sized double emulsions, created using poly(dimethylsiloxane) (PDMS)-based membranes[1]. These nanocultures, equipped with semi-permeable membranes and magnetic nanoparticles (MNPs), aim to provide a solution to the long-standing challenge of cultivating and retrieving previously unculturable microbial species from environmental samples [2]. We hypothesize that the MNCs can be incubated in simulated natural environments, enabling the study of microbial growth dynamics, and the retrieval of MNCs after incubation without substrate confinement, preserving natural diffusion dynamics [3]. The study addresses the gap in our understanding of unculturable microorganisms, offering potential applications in drug discovery, high-throughput screening, and microbial ecology across various environmental conditions. **Methods:** Transmission and Scanning Electron Microscopy were performed to characterize the magnetic nanoparticles (MNPs) and generated nanocultures. Magnetic polymer membranes (MPMs) were characterized by measuring the water contact angles, mechanical testing, EDS point spectra, and elemental maps of the bulk MPMs. A vibrating sample magnetometer was employed to obtain the magnetic susceptibility of the functionalized polymers. Fluorescence intensity via confocal imaging was utilized to quantify bacterial growth in the NCs. Computational tools were used to obtain average size distributions of MNPs embedded within MPMs. *Escherichia coli* strains, including Enhanced Green/Red Fluorescent Protein-tagged variants, were encapsulated in MNCs and placed under silica beads to mimic soil-like conditions. The study demonstrates successful retrieval of MNCs using magnetic actuation without hindering microbial growth. Computational methods were employed to calculate the average velocity of the NCs. These novel nanocultures offer a means to study microbial growth under various environmental conditions while enabling retrieval for downstream omics studies. **Results and Implications:** Magnetic susceptibility as a function of field was measured by magnetometry and found to be 19×10^{-7}, 12.6×10^{-7}, and 12.0×10^{-7} emu/Oe for 10 nm, 20 nm, and 5 nm MNPs respectively. The average velocity of the MNCs under a magnetic field gradient of 1.48 T, was approximately 80 µm/s, 200 µm/s, and 130 µm/s for 5 nm, 10 nm, and 20 nm diameter MNPs, respectively. Simulations of magnetophoresis are in reasonable agreement with the main features of the experimental measurements. The study lays the foundation for further exploration of environmental microbiomes and the discovery of previously unknown microbial species and associated beneficial secondary metabolites. Future research questions include optimizing MNCs design for different environmental settings, investigating cross-kingdom species interactions within nanocultures, and expanding applications in drug discovery and therapeutic microbiology.

References:

[1] **Usman, Huda** et al. (2021) Design of a well-defined poly(dimethylsiloxane)-based microbial Nanoculture system. Materials Today Communications doi:10.1016/j.mtcomm.2021.102185.

[2] Manimaran, N.H.*, **Usman, Huda*** et al. (2020) Developing a functional poly(dimethylsiloxane)-based microbial Nanoculture system using dimethylallylamine, ACS Applied Materials and Interfaces, 12(45), pp. 50581–50591. doi:10.1021/acsami.0c11875.

[3] **Usman. Huda** et al (In Preparation), Designing Magnetically Responsive Nanoculture System.

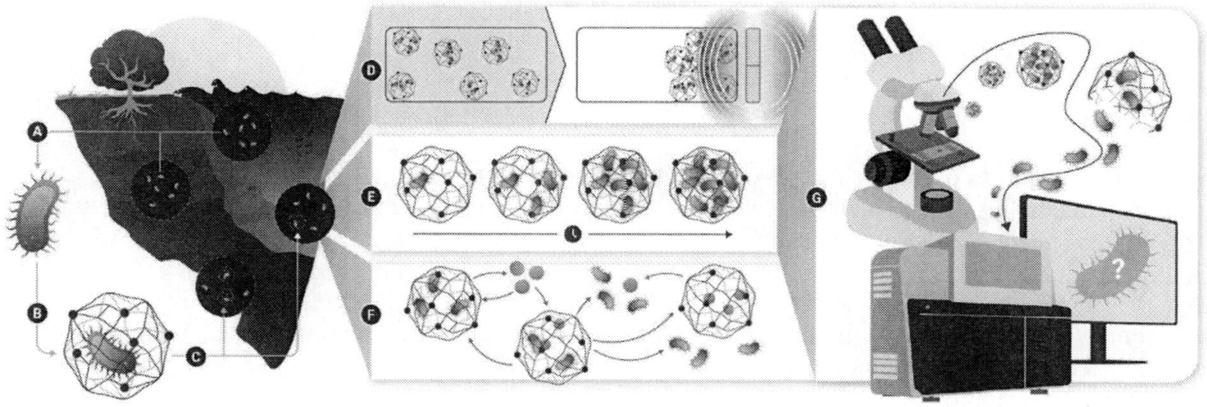
A
B
C
D
E
F
G
?

339a Determining Spinnability of Carbon Nanotube Dispersions in Wet Spinning Via Shear Viscosity of Dispersions

Jaegyun Im, Jaegeun Lee

Unprecedent properties of nanomaterials can be utilized by manufacturing them into macroscopic structures. Assemblies of nanomaterials have been widely manufactured using the liquid phase, including both a molten state and in form of a solution. Particularly, the complex phase and flow behaviors of solutions facilitate various approaches to understand and control the manufacturing process.

Wet spinning, a representative solution processing, is the fiber spinning process where the spinning dope passes through a narrow spinning line and solidifies into fiber. Especially, when the spinning dope flows through a narrow spinning line, the flow occurs under the high shear-rate regime. The dynamic behavior of materials in this regime affects the structure of the as-spun fiber which determines the success of the fiber formation, becoming a matter of interest in the field of wet spinning.

In the wet spinning of polymer fibers, the spinnability of dopes, the ability of dopes to spin fibers continuously, has been determined using their shear viscosity. For the quantitative evaluation of the flow behavior from the shear viscosity, the power law viscosity model has been employed due to its simplicity. The power law viscosity model, known as the Ostwald-de-Waele relationship, describes the shear viscosity η [Pa·s] as a function of the shear rate γ [s^{-1}] by the following equation:

$$\eta = m\gamma^{n-1}$$

where is the consistency index m [Pa·s^n], and n is the power law index [dimensionless]. We can quantify the flow behavior of fluids through . When n=1, the fluid exhibits the Newtonian flow behavior which has a constant shear viscosity. For n>1, the fluid exhibits the shear-thickening behavior, while the fluid with exhibits the shear-thinning behavior. As n approaches to 0, the shear-thinning behavior becomes stronger. The evaluation of the spinnability through the quantified flow behavior of dopes enables to discover appropriate dope conditions, advancing the wet spinning.

Among promising materials applied in the wet spinning, carbon nanotubes (CNTs) have been frequently processed in macroscopic-fiber form. The wet spinning of CNT fibers greatly advanced by the solution processing for homogeneous CNT dispersions based on the superacid and surfactant system. However, unlike the wet spinning of polymer dopes, the flow behavior of CNT dispersions within the spinning line has not been investigated. The rod-like structure and large aspect ratio of CNTs similar to the long-chain characteristics of polymers enabled us to apply the methods for polymer dopes into CNT dispersions. For instance, the phase behavior of CNT dispersions and characteristics of CNTs can be analyzed using the Onsager theory which is based on polymer thermodynamics. The similarity between polymers and CNTs inspired us to determine the spinnability of CNT dispersion in wet spinning using the power law viscosity model as used in polymers. Herein, we presented the correlation between the spinnability and the flow behavior of CNT dispersions through the power law viscosity model.

We investigated the flow behavior of CNT dispersions at various concentrations using three commonly used single-walled CNT products (eDips, Tuball, SG101) in the wet spinning of CNT fibers. The surfactant-based aqueous CNT dispersions were obtained after the ultrasonication. Prior to evaluate the shear-thinning behavior between the dispersions through the power law viscosity model, we investigated the shear-rate range which can represent the flow behavior of the dispersion within during the wet spinning, enabling proper fitting of the model. The shear rate γ in the spinning line was estimated using the following equation, assuming laminar flow:

$$\gamma=4Q/\pi r^3$$

where Q is the volume rate of the spinning rate, and r is the radius of the spinning needle. In our wet spinning, the shear rate within the spinning line was estimated as 787.33 s^{-1}. The shear-rate range from 100 to 400 s^{-1} and from 200 to 800 s^{-1} which are close or contains the estimated spinning shear rate were investigated for the power law viscosity model. We quantitatively compared the suitability of the fitting ranges using the normalized difference between the expected and experimental viscosity values at the estimated shear rate of spinning. Despite similar R^2 values of the fittings, the normalized differences where the fitted range was from 100 to 400 s^{-1} were greater than that were fitted within the range of 200 to 800 s^{-1} for all CNT dispersions. This suggested that the fitting range which includes the spinning shear rate more accurately predicts the shear viscosity at the spinning shear rate than other. In this work, we determined the spinnability of the dispersion using the power law index from the power law viscosity model fitted within the shear rate range of 200 to 800 s^{-1}.

We defined the spinnability as the ability to spin fibers continuously with measurable properties. The spinnability was classified into the 'good' spinnability when the fiber spinning is continuous and the fiber properties are measurable, and the 'bad' spinnability wherein the fiber spinning is discontinuous and the fiber properties are unmeasurable. eDips, Tuball, and SG101 had the good spinnability when was lower than 0.418 (0.16 wt.%), 0.438 (0.35 wt.%), and 0.738 (0.48 wt.%), respectively. Although CNT products exhibited different ranges of the spinnable , the spinnability and fiber properties were increased as approached to 0. In other words, the spinnability of the dispersion was proportional to the shear-thinning behavior of the CNT dispersion.

When the CNT dispersion flows, CNTs encounter each other, resulting in the friction forces between them. The friction forces induce the disentanglement and reorientation of CNTs towards to the shear direction, facilitating the flow of the dispersion in the shear direction. As a result, the shear viscosity of the dispersion decreases, which known as shear-thinning behavior. The weak shear-thinning behavior was observed at the dispersion with low CNT concentration which had large free space within the dispersion. Based on polarized optical microscopic images and the magnitude of shear viscosity, we confirmed that the dispersion at low concentration had large free space. When the free space within the dispersion is large, the amount of contact between CNTs during the flow will be minimized, decreasing the possibility of collision between them. The friction forces will not be sufficient to induce the disentanglement and reorientation of CNTs, exhibiting the weak shear-thinning behavior. In the wet spinning, as CNTs within the dispersion maintain an isotropic conformation even at the spinning shear rate, the resulting CNT fiber will exhibit a low contact area between CNTs within the fiber. Consequently, the 'bad' spinnability was observed in this dispersion.

On the other hand, the strong shear-thinning behavior was observed at the dispersion with high CNT concentration which had small free space within the dispersion. The increase of the birefringence and shear viscosity at high CNT concentration indicates the small free space within the high CNT concentration dispersion. The amount of contact between CNTs will large enough to induce the friction forces resulting in the disentanglement and reorientation of CNTs. The CNT fiber will have a large contact area between CNTs within the fiber, exhibiting the 'good' spinnability.

The correlation between the spinnability and the shear-thinning behavior of CNT dispersion arises from the fiber integrity which is derived from the strong van der Waals forces between CNTs. If CNTs are aligned along the fiber axis, the total amount of van der Waals forces between CNTs within the fiber will be maximized as the contact area between CNTs increases. We hypothesized that CNT fibers from CNT dispersions which has the 'good' spinnability have the large contact area between CNTs within the fiber. In contrast, the dispersion exhibiting the weak shear-thinning behavior are not able to form fibers due to the small contact area between CNTs which derives from the isotropic conformation of CNTs within the fiber.

We can extend our hypothesis to other materials with similar fiber integrity as CNTs. When the fiber integrity of the material is governed by the strong van der Waals force of the graphitic carbon structure, the spinnability will be maximized as the dispersion close to the ideal shear-thinning behavior. Similar to

CNT, graphene oxide (GO) has the strong van der Waals force on its graphitic plane. GO fibers maintain their structure through the strong van der Waals forces between adjacent layers. The spinnability of GO dispersions in the wet spinning was found to be proportional to the shear-thinning behavior of the dispersions. Therefore, in the wet spinning of the carbon materials with graphitic structure, we suggest that the spinnability of the dispersion can be determined through the degree of the shear-thinning behavior of the dispersion.

In summary, we quantitatively investigated the correlation between the shear-thinning behavior and the spinnability of CNT dispersion through the power law index which approaches to 0 if the dispersion has stronger shear-thinning behavior. The dispersions were more spinnable as approached to 0, suggests that the spinnability of CNT dispersions is proportional to the shear-thinning behavior of the dispersion. Our work presents a simple method to determine the spinnability of CNT dispersion using the shear viscosity of CNT dispersion.

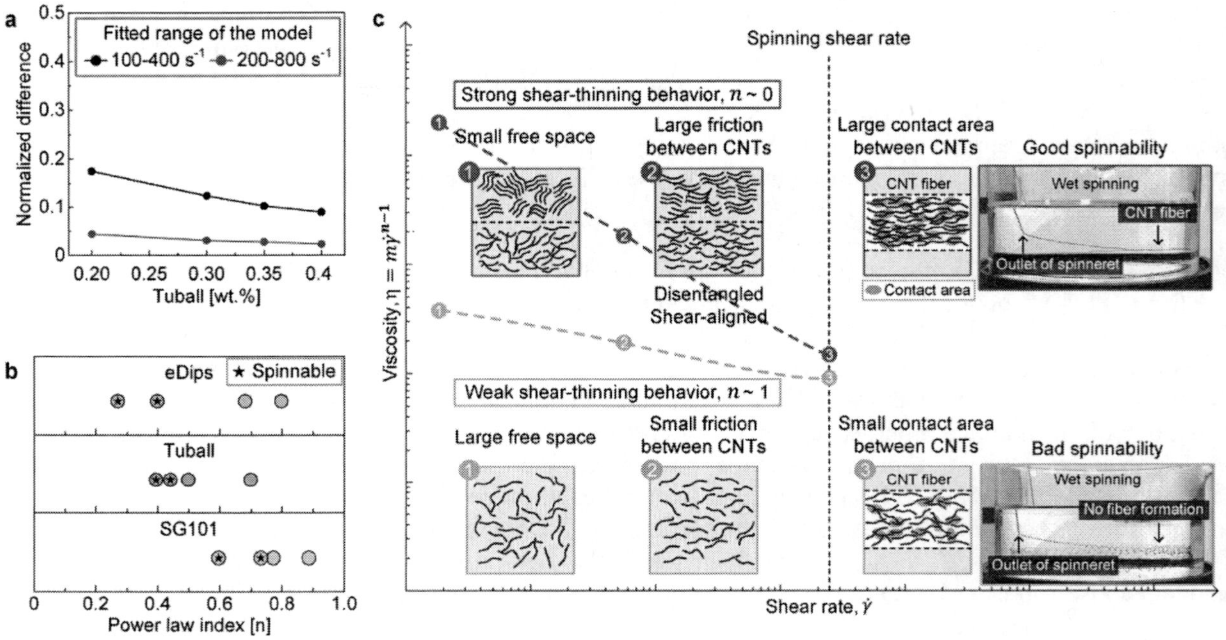

Figure. (a) The normalized difference between the expected and experimental viscosity values, (b) the power law index of CNT dispersions, (c) the shear-thinning behavior and spinnability of CNT dispersions

339b Insight into Selective Binding of Carbon Dioxide on Carbon Quantum Dots through Molecular Dynamics and Monte Carlo Simulations at the Atomic Level

Mohsen Samandari, Michael Broud, David Harper, David Keffer

To effectively tackle the challenges posed by global climate change, a practical approach involves capturing CO_2 directly at its source; thus, utilizing an abundant, sustainable, and cost-effective adsorbent with high selectivity is crucial. This study leverages established synthesis methods to create adsorbents from Kraft softwood lignin (KSL), a biobased material that meets these criteria and is recyclable, while also enabling the recovery of adsorbed gases. KSL, a coproduct produced in large quantities at pulp and paper mills worldwide, serves as the foundation for this research.

Enhancing the internal surface of the adsorbents using carbon quantum dots (CQDs), which were derived from lignin themselves, was found to be an effective method to increase adsorption capacity and selectivity of CO_2 over other gas molecules. CQDs facilitate the selective adsorption of CO_2 gas molecules driven by electrostatic interactions, taking advantage of the unique quadrupole moments exhibited by CO_2, N_2, and O_2. This manipulation of charge distribution on CQDs can be achieved through methods such as nitrogen or oxygen doping during synthesis, or by introducing amine, carboxyl, or hydroxyl groups within the aromatic structure or at the edges.

To evaluate the binding strength, adsorption isotherms and selectivity of CQDs towards CO_2, N_2, and O_2, a combination of classical molecular dynamics simulations and Grand Canonical Monte Carlo simulations were employed across various carbon substrate materials at different temperatures. Statistical analysis of CQD-gas interactions was used to quantify adsorption selectivity and capacity. The results demonstrate the potential to modify binding strength and selectivity for CO_2 over N_2, and O_2 by controlling the charge distribution of CQDs on a carbon substrate. Ongoing efforts are focused on optimizing the atomic structure and dimensions of CQDs, as well as the extent of doping and functionalization, to achieve a viable large-scale adsorbent solution for CO_2 capture.

339c Single-Molecule Modifications of ssDNA-Wrapped SWCNT By Magnetic Beads-Assisted Sorting

Shoichi Nishitani, Deverin Smith, Zirun Liang, Markita Landry

Single-walled carbon nanotubes (SWCNTs) offer significant potential for probing molecular interactions at single-molecule resolution, owing to their unique optical/electronic properties and their one-dimensional structure. However, achieving precise SWCNT surface modifications has been challenging, primarily due to the difficulty in manipulating chemical reactions at the single-molecule scale. In this study, we propose a facile, magnetic beads-assisted sorting strategy to achieve single-molecule modifications of SWCNTs.

Our methodology encompasses a two-step process: 1- Synthesizing SWCNT dispersion with a mixture of unmodified ssDNA and biotin conjugated ssDNA (ssDNA-Bt). With a sufficiently low ssDNA-Bt/ssDNA ratio in the starting material, we predict SWCNTs that have ssDNA-Bt, will only have one ssDNA-Bt at a distinct probability; thus, a single biotinylation on a SWCNT. 2- These singly-biotinylated SWCNTs are selectively isolated using streptavidin-coated magnetic beads. To overcome the strong biotin-streptavidin binding during recovery, we employ desthiobiotin (dtBt), a biotin analogue with reduced affinity for streptavidin, facilitating the release of SWCNTs through an affinity-based exchange. Finally, a single biomolecule can be conjugated selectively to a SWCNT through the ssDNA-dtBt bond either by using a streptavidin-assisted conjugation or by including conjugation chemistry on a ssDNA-dtBt strand.

As a proof-of-concept, we prepared $(GT)_{15}$-dtBt/$(GT)_{15}$-SWCNT with a starting mixing ratio of as low as 1/1000. The synthesized SWCNT sample was then incubated with streptavidin-coated magnetic beads for 1 h, followed by the addition of biotin to recover the captured biotinylated SWCNTs. The recovery of SWCNT was characterized by near-infrared (NIR) fluorescence microscopy. Accordingly, 80 % of the dtBt-functionalized SWCNTs were selectively recovered. Notably, recovery from the control samples [$(GT)_{15}$-Bt/$(GT)_{15}$-SWCNT or $(GT)_{15}$-SWCNT] was negligible, highlighting the ability of our strategy to enable single molecule SWCNT functionalization with high efficiency. We further quantified the functionalization sites by atomic force microscopy and an enzymatic activity assay. Thus, we herein developed a facile strategy of single-molecule functionalization of SWCNTs, a method which may be generalized for broader applications in SWCNT based sensing and sorting, including chiral separations and single-defect engineering.

339d An Autonomous, Closed Loop Research System for Scalable Carbon Nanotube Synthesis

Arthur Sloan, Morgen Smith, Robert Waelder, John Bulmer, Benji Maruyama

Autonomous experimentation systems are becoming increasingly prevalent in the field of materials science as a means to accelerate discovery and development of new materials. Such systems combine three key capabilities: robotic execution of experiments, in situ or in line analysis techniques, and the use of machine learning to plan follow-up experiments based on the collected data. By iterating through this experiment-analyze-plan cycle, autonomous research systems can pursue a specified goal an order of magnitude more efficiently than traditional design of experiment techniques, as benchmarked by the number of required experiments, while also consuming less material and researcher time.[1] These productivity gains have been demonstrated across diverse materials science problems, ranging from carbon nanotube (CNT) synthesis to topological optimization of 3D printed components, to the exploration of material phase transitions.[2,3,4]

Previous closed-loop research on CNT synthesis from our group has used the autonomous research system (ARES) to investigate on the surface growth of CNT in a cold-wall chemical vapor deposition (CVD) reactor using laser heating.[2,5,6] While this system has provided valuable insight into factors affecting the growth rate of CNTs, the results are not directly applicable to the scale of up controlled CNT production to commercially viable levels. To address this shortcoming we demonstrate the operation of a new ARES systems based around the floating catalyst chemical vapor deposition (FCCVD) process for CNT synthesis. FCCVD is a substrateless process that can be operated continuously and is thus a promising route for the scaled production of high-quality single or few-walled CNTs.[7]

This floating catalyst ARES (FC-ARES) consists of a benchtop FCCVD reactor with computer control of reactor temperature and the flow rates of up to three gas phase and four liquid phase feedstock species. CNT samples are collected on a reel-to-reel tape at the reactor outlet.[8] Samples deposited on this tape are advanced through one or more analysis stations to collect experimental metrics such as the presence and distribution of CNT radial breathing modes or the yield of the experiment. Experimental control, data collection, and experimental planning are centralized using our open-source ARES OS™ software. Building upon our previous work in the in cold-wall CVD reactor ARES, we utilize a machine learning planner based on the expected improvement decision policy to demonstrate the usefulness of the FC-ARES in optimizing both CNT yield and controlling SWCNT diameter.[5] Additionally, we compare the results of the autonomous experiments to traditionally planned experimental campaigns as a benchmark.

References

- Stach, E. et al. *Matter* **4**, 2702–2726 (2021).
- Nikolaev, P. et al. *npj Comput Mater* **2**, 16031 (2016).
- Gongora, A. E. et al. *Science Advances* **6**, eaaz1708 (2020).
- Kusne, A. G. et al. *Nat Commun* **11**, 5966 (2020).
- Rao, R. et al. *npj Comput Mater* **7**, 157 (2021).
- Waelder, R. et al. *Carbon* (In Review)
- Weller, L. et al. *Carbon* **146,** 789–812 (2019).
- Rao, R., and Maruyama, B. *S. Patent No. 10,994,990.* (2021).

339e Chirality-Specific Protein Adsorption on Single-Walled Carbon Nanotubes

Robert Nißler, Rebecca Pinals, Sibylle Pfammatter, Han Li, Benjamin Flavel, Inge Herrmann

Semiconducting single-walled carbon nanotubes (SWCNTs) serve as versatile near-infrared (NIR) fluorophores, offering a substrate for both noncovalent and covalent modification to create sensors that display modulated fluorescence upon target analyte interaction. Such sensors can be applied in basic biological research and biomedical diagnostics. The fluorescence emission of SWCNTs depends on their lattice structure, or so-called chirality (*n,m*), enabling the development of multi-color biosensors, including those for cancer biomarker detection in human plasma. However, when exposed to complex body fluids, such biosensors are influenced by protein corona formation, a process in which proteins from the surrounding fluid absorb to the nanotube surface. Profiling protein adsorption on mixed-chirality SWCNTs has revealed the composition and driving forces of protein corona formation, leaving chirality-specific interactions unexplored. Building on this work[1], we hypothesize that the nanotube chirality, and thus diameter and surface curvature, play a key role in determining protein-nanotube interactions and corona formation. Yet, chirality-specific interactions remain difficult to study due to the demanding production of sorted SWCNT material and hence limited quantities for proteomic investigations. Here, we employ mass spectrometry analysis adapted from single-cell-proteomics to study the protein corona composition on chirality-pure SWCNTs. Aqueous two-phase extraction facilitates the reliable sorting of single-color probes. Functionalizing these single-chirality SWCNTs with biopolymers such as ssDNA then creates a biosensor. Initially, we analyze protein adsorption from human plasma on a single chirality, specifically (6,5)-SWCNTs, modified with various oligonucleotide (ssDNA) sequences that display different sensing and nanotube-binding properties, to assess the impact of surface chemistry on the resultant protein corona profile. Subsequently, we investigate how protein adsorption is influenced by SWCNT chirality bearing the same surface modification. Notably, our protocol enables the recycling of precious chirality-sorted SWCNT material. We establish a foundation for correlating NIR fluorescence sensing of disease states with protein biomarker adsorption. Overall, our study unveils how SWCNT surface chemistry and chirality affect protein adsorption, enabling future targeted biosensing approaches.

1. Pinals, R. L. et al. Quantitative Protein Corona Composition and Dynamics on Carbon Nanotubes in Biological Environments. Angew. Chemie - Int. Ed. 59, 23668–23677 (2020).

339f Single-Chirality Carbon Nanotubes for NIR Five-Dimensional Optoacoustic Imaging

Robert Nißler, Daniil Nozdriukhin, Isabel Klessascheck, Han Li, Benjamin Flavel, Daniel Razansky, Inge Herrmann, Xosé Luís Deán-Ben

Optoacoustic (OA, photoacoustic) tomography offers a unique five-dimensional imaging capacity, combining real-time, multi-spectral (multi-wavelength), and three-dimensional imaging of spectrally-distinct substances. Contrast agents with defined absorption peaks within the near-infrared window (NIR, 700-1350 nm) enable deep-tissue visualization of dynamic biological processes in vivo. However, spectral coloring effects associated with wavelength-dependent light attenuation hinder simultaneous differentiation of multiple agents, limiting the OA multiplexing capacity and restricting the full potential of this modality. One class of nanomaterials that could help overcome these limitations are semiconducting single-walled carbon nanotubes (SWCNTs), which exhibit a lattice-dependent bandgap structure known as chirality (n,m). In this study, we introduce chirality-separated carbon nanotubes as OA contrast agents characterized by narrow and tunable absorption peaks within the NIR. These sorted SWCNTs were obtained by aqueous two-phase extraction (ATPE) and later surface-modified with macromolecules to enhance their biocompatibility. The nanotubes' unique optical features enable unambiguous differentiation of multiple types of nanotubes amidst strong background signals from blood and other biological tissues, while exhibiting excellent photostability. High-frame-rate in vivo imaging of a nanotube bolus injected into the bloodstream was accomplished with only three optical wavelengths, while nanotube-coated microparticles could be individually detected and tracked. These results pave the way for innovative approaches in perfusion assessment, super-resolution imaging, and targeted drug delivery.

339g Competing Mechanisms of in-Plane Thermal Rectification in Semi-Stochastic Grain-Size Graded Graphene Structures

Simanta Lakhar, Raghavan Ranganathan

Thermal rectifiers, analogous to their electronic counterpart, are devices that preferentially conduct heat in one direction, and can potentially open up novel technological frontiers of applications such as, thermoelectric power generation using thermal diode bridge, thermal logic circuits, and advanced thermal management devices for micro-electronics and atomic clocks, if they can be realized with adequate efficiencies[1]. Compared to bulk materials, where asymmetric heat flow characteristics can be achieved through few specific mechanisms, two dimensional (2D) materials could offer a wide range of strategies or designs for generating and tuning thermal rectification (TR) behavior, which can be attributed to mechanisms such as difference in the ease of heat flow in the two opposite directions due to resonance or mismatch in the regional phonon spectra for either case, respectively, asymmetric scattering of propagating phonons depending on the direction of heat flow, and difference in the temperature-dependence of thermal conductivity in the two sides of the rectifier device[2]. Realization of practical thermal rectifiers relies significantly on the sound understanding of the underlying mechanisms of asymmetric heat transport so that the different strategies can be tuned and potentially combined for increased effectiveness. However, the in-plane thermal rectification in 2D materials like graphene having directional gradients of grain sizes, which could potentially exhibit an interplay of multiple mechanisms including varying phonon localizations throughout the graded structure, asymmetric phonon scattering and asymmetric temperature dependence of thermal conductivity, have remained elusive. In fact, understanding the heat transport mechanisms in polycrystalline graphene, which are more practical to synthesize than large scale single crystal graphene, could potentially be even more useful due to the high in-plane thermal conductivity of graphene as compared to other 3D designs of thermal rectifiers, like the graphene nanotubes and cross-plane graphene interface[2], and allow a unique opportunity to combine with other defects and designs for effective optimization of the thermal rectification property.

In this work [3], we investigated the thermal rectification behavior in periodic atomistic models of polycrystalline graphene whose grain arrangements were generated using different probability distribution functions to have different gradient grain size distributions along the heat flow direction in the graphene layer [4]. We employed the centroid Voronoi tessellation technique to generate realistic grain boundary structures for graphene[4]. The reverse non-equilibrium molecular dynamics simulations[5], which is a key tool in this area in order to understand the asymmetric heat transport phenomena, was used to calculate the thermal conductivity and rectification of the structures equilibrated at different stress and temperature conditions. Additionally, detailed phonon mapping was carried out to understand the underlying mechanisms based on the fluctuation-dissipation theory [6] in order to explain the competitive interplay between two transport mechanisms, namely, (1) propagating phonon coupling and (2) temperature-dependence of thermal conductivity, that in turn, determine the degree of asymmetric heat flow in graded polycrystalline graphene. A graphical overview of this work is shown in Figure 1. Based on a data-driven analysis, we propose the ideal parameters of a gradient grain-size structural design in graphene which can lead to an optimal thermal rectification device.

References

[1] M. Wong, C. Tso, T. Ho, and H. Lee, 'A review of state of the art thermal diodes and their potential applications,' *Int. J. Heat Mass Transfer,* vol. 164, p. 120607, 2021.

[2] S. Zhao, Y. Zhou, and H. Wang, 'Review of thermal rectification experiments and theoretical calculations in 2D materials,' *Int. J. Heat Mass Transfer,* vol. 195, p. 123218, 2022.

[3] S. Lahkar and R. Ranganathan, 'Competing mechanisms govern the thermal rectification behavior in semi-stochastic polycrystalline graphene with graded grain-size distribution'. *Carbon, 218,* 118638, 2024.

[4] C. Ophus, A. Shekhawat, H. Rasool, and A. Zettl, 'Large-scale experimental and theoretical study of graphene grain boundary structures,' *Phys. Rev. B,* vol. 92, no. 20, p. 205402, 2015.

[5] Q.-X. Pei, Y.-W. Zhang, Z.-D. Sha, and V. B. Shenoy, 'Carbon isotope doping induced interfacial thermal resistance and thermal rectification in graphene,' *Appl. Phys. Lett.,* vol. 100, no. 10, p. 101901, 2012.

[6] L. T. Kong, 'Phonon dispersion measured directly from molecular dynamics simulations,' *Comput. Phys. Commun.,* vol. 182, no. 10, pp. 2201-2207, 2011.

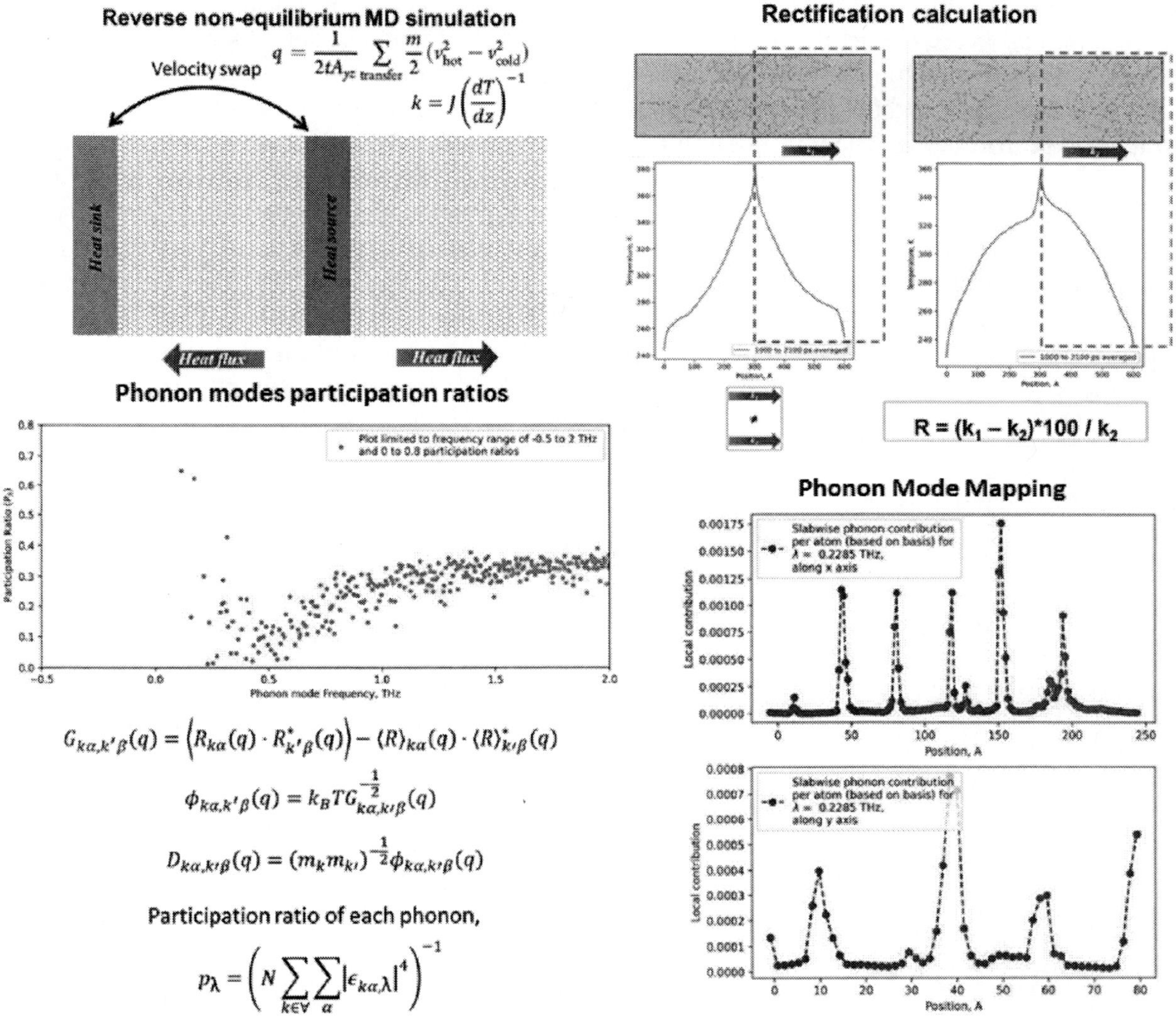

301a Experimentally Guided Time Series Models for Carbon Nanofibers Mixed Matrix Membranes for Enhanced Water/Salt Permselectivity

Saketh Merugu

Processes involving liquid-vapor phase change mechanisms such as membrane distillation rely on solid-liquid interface with superior interfacial characteristics such as porosity, wettability, surface tension and adsorption. The talk will outline a systematic approach to designing efficient membranes for separation processes by using predictive models to forecast water vapor flux and salt rejection rate datasets (N= 434), thereby predicting long-term membrane stability and performance. The focus will be on the influence of carbon nanofibers mixed matrix membranes on membrane properties and desalination performance. The introduction of just 1 wt% Cu/CNF resulted in enhanced contact angle hysteresis and hierarchical morphologies, providing a pathway for water vapor escape and recording a 64% increase in water vapor flux compared to PVDF membranes, while maintaining a 99.9% salt rejection rate. The study highlights the superior accuracy of statistical time series models in predicting the performance of the membrane distillation process, while noting the limitations of deep learning algorithms in this context. The research suggests further exploration of alternative validation methods and additional models to enhance prediction accuracy. The findings could lead to more efficient membrane distillation processes and provide valuable insights for future research in developing precise models for predicting membrane performance.

301b Protein Derived Hierarchically Porous Graphitic Aerogels: Unraveling Formation Mechanism and Their Application in EMI Shielding

M. Shaharyar Wani, Yusuf Jimoh, Paul Prucnal, Craig B. Arnold

Hierarchically porous graphitic sheet-based aerogels (HGAs) are ultra-low density materials that have garnered significant attention across various fields, including catalysis, electromagnetic shielding, energy storage/conversion, water purification, gas storage/separation, and biomedical applications, due to their exceptional electrical, electronic, electrochemical, and mechanical properties. However, current fabrication methods involve multiple steps, generate harsh chemical wastes, and lack covalent integration of graphitic sheets, which compromises their properties and scalability while rendering them unsustainable [1].

A particularly promising solution for enabling sustainable, bottom-up synthesis of such carbon structures is the use of naturally available protein precursors. This is due to their innate capability to form supramolecular aggregates [1]. To effectively implement proteins for developing complex graphitic aerogel self-assembly, it is essential to elucidate the underlying transformation mechanisms from precursor to the final hierarchically porous morphology. Furthermore, unraveling the intricate processing-structure-property relationships that dictate material optimization is critical for enabling the scalable production of high-performance and sustainable HGAs tailored for emerging applications.

This work introduces a novel self-foaming mechanism that enables the green and scalable synthesis of HGAs using naturally available protein precursors. Using in-situ characterization techniques, including optical microscopy complemented by time-resolved thermal and mass spectrometry, we elucidate the structural and chemical evolutions during protein transformation and understand the intricate foaming process. We found that controlled heating of protein precursors during a single-step pyrolysis induces intrinsic foaming via initial softening, gas evolution, and graphitization [2]. The liberated gases facilitate bubble expansion and coalescence, while the increasing temperature immobilizes and graphitizes the foamed structure, resulting in a hierarchically porous graphitic aerogel composed of a seamlessly integrated framework of sheets and fibers.

Building on our understanding of the formation mechanism, a systematic exploration of processing-structure-property relationships was conducted to tailor critical material performance metrics, including density, porosity, and electrical conductivity, and their impact on EMI shielding applications was investigated [3]. The bio-derived HGAs exhibited a specific shielding effectiveness per unit thickness (SSE/t) of over 16,400 $dBcm^2/g$, governed by an absorption-dominated mechanism, which is significantly higher than previously reported carbon/graphene aerogels.

In summary, this simple yet effective synthesis strategy highlights the potential of protein precursors for engineering HGAs for diverse applications, such as high-performance EMI shielding. Elucidating this multidimensional parameter space facilitates unraveling the intricate interplay between tunable synthesis variables and resultant aerogel properties and structure. This enables the deliberate modulation of microstructural features for customized bulk functionality. This study provides guiding design rules for replicating the sustainable, bottom-up synthesis of high-performance HGAs in a facile one-step process, paving the way for their scalable production in next-generation technologies.

References:

[1] Ozden, S. *et al.* Egg protein derived ultralightweight hybrid monolithic aerogel for water purification. *Mater. Today* **59**, 46–55 (2022).

[2] Wani, M. S., *et al.*, Self-Foaming Morphogenesis of Hierarchically Porous Graphitic Aerogel (submitted)

[3] Wani, M. S., *et al.*, Protein Derived Hierarchically Porous Graphitic Aerogel for Electromagnetic

Shielding Interference (in preparation)

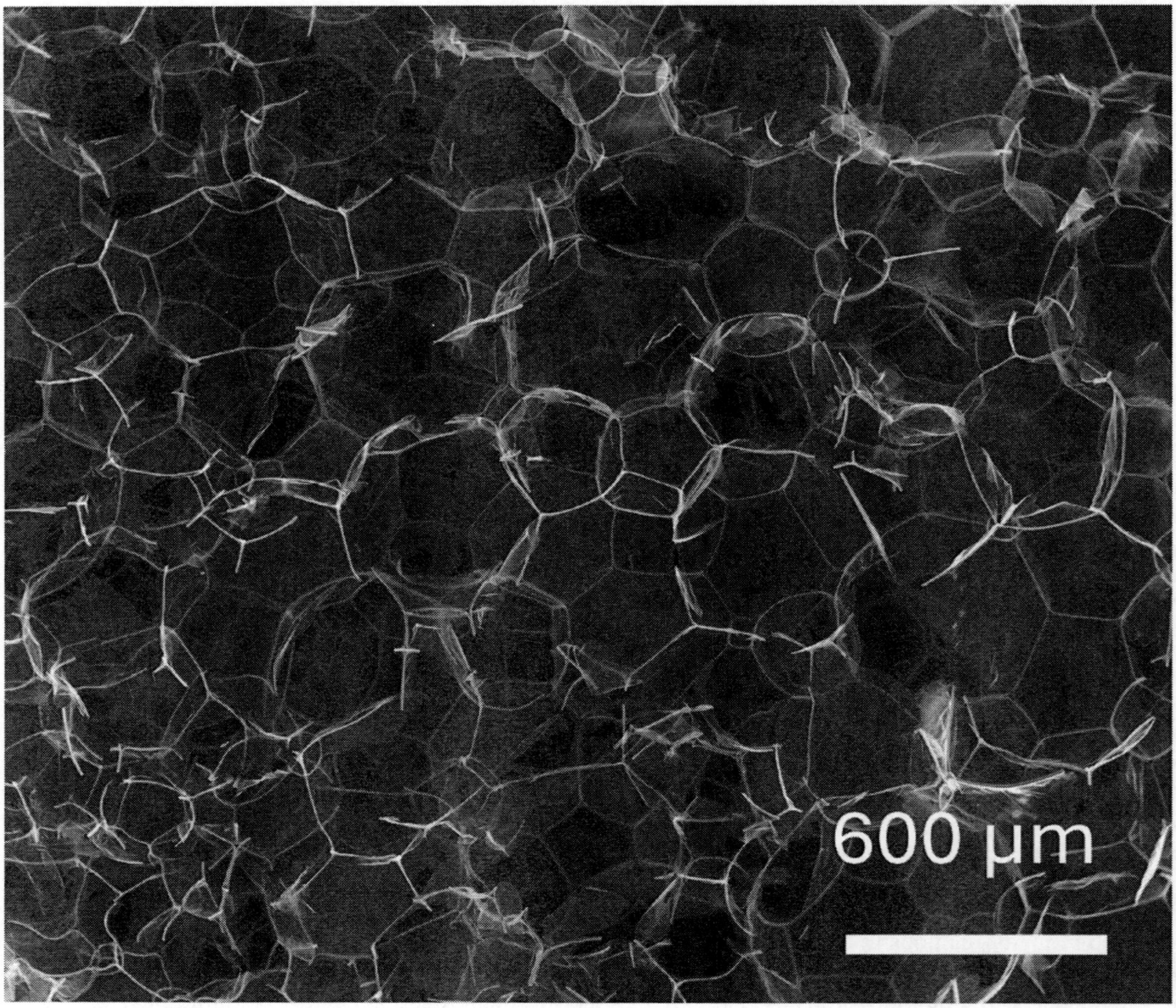

301c Enhanced Lithium Recovery through Innovative 2D Nanocomposites and Membrane Technologies: A Pathway to Sustainability

Nour Abdelrahman, Seunghyun Hong, Hassan Arafat, Daniel Choi, Faisal Al Marzooqi

Given the growing need for lithium due to its essential role in electric vehicles and renewable energy storage, our study addresses the crucial task of effectively extracting lithium from a variety of water sources, including brines, wastewater, and seawater. Our aim is to promote a more eco-friendly life cycle for lithium-ion batteries through the introduction of innovative technological solutions, focusing on adsorption and membrane-based separation methods.

Our research highlights the use of two-dimensional (2D) nanocomposite materials known for their remarkable ability to selectively and efficiently capture lithium ions. We have developed a pioneering adsorbent composed of sulfonated graphene oxide, MXene, ionic liquid, and alginate, which showcases superior lithium adsorption and removal capabilities, achieving more than 89% efficiency in lithium ion removal.

In parallel, we explore the potential of membrane-based separation as an environmentally friendly alternative for lithium extraction. We present a new type of membrane made from vertically aligned, functionalized transition metal carbides and nitrides (MXene). These membranes stand out for their fast lithium ion transport and ability to effectively separate lithium ions from other ions, based on their size and different affinities. Our modified membranes have shown exceptional performance, with a Li^+/Mg^{2+} selectivity ratio exceeding 104, far outperforming conventional membranes, and achieving over 94% rejection of $MgCl_2$.

By combining 2D nanocomposite adsorption with advanced membrane separation technology, we offer a comprehensive solution for lithium recovery. Implementing these advanced technologies not only reduces the environmental impact of lithium mining but also advances lithium recovery techniques. This approach strengthens the sustainability of the lithium-ion battery industry, ensuring its future in light of limited lithium resources and growing environmental concerns.

301d Investigating Adsorption of Mono/Oligosaccharides on Carbon Nanotubes Using Molecular Simulations

Parisa Farzeen, Sanket Deshmukh

Carbon nanotubes (CNTs) have potential applications in areas of energy, tribology, nanoelectronics and biomedicine because of their high thermal stability, high thermal conductivity, current-carrying capabilities, and other electrical, structural, and physicochemical properties. Despite these exceptional properties, due to their poor water solubility and toxicity including oxidative stress, inflammatory responses, and more, CNTs have not been widely explored in the medical field. To overcome these challenges, CNTs have been decorated using various water-soluble moieties via chemical treatment. However, chemical treatment could be challenging because such treatment can significantly alter the chemical and physical properties of the CNTs. A possible solution is to employ mono- and di-saccharides, and their derivatives to decorate CNTs *via* adsorption, which could potentially enhance their biocompatibility and water solubility due to the presence of multiple hydroxyl groups. Here, we have employed all-atom and coarse-grained molecular dynamics (MD) simulations to study the adsorption of these molecules on CNTs by calculating free energies using the potential of mean force (PMF) method. Specifically, we investigated the effect of CNT curvatures and single- vs multi-wall CNTs on the adsorption of carbohydrate molecules. This study paves the path for selecting carbohydrates for decorating CNTs for biomedical applications such as drug carriers.

607a Combining Single-Crystal Oxides and First Principles Computations to Understand Electro and Photocatalysis on Oxides

Geoffroy Hautier

Hydrogen production from water through electro- or photo-catalysis is an attractive path towards the decarbonation of our energy system. However, the typical catalysts used for the water splitting reactions have important limitations in stability, activity or earth-abundance. An atomistic understanding of the catalysts surface and its reactivity would enable a better design of new catalysts but has been challenging to achieve.

With the advances in single crystal growth especially of oxides, very high-quality surfaces can be obtained offering a model platform to understand catalysis. I will report on the efforts to understand the oxygen evolution reaction on oxides (OER) under the Center for Electrochemical Dynamics And Reactions on Surfaces (CEDARs), an Energy Frontier Research Center (EFRC) from the Department of Energy (DOE). The first part of my talk will focus on RuO2 and IrO2 rutile has they are among the most attractive OER catalysts. I will show how theory and experiment on molecular beam epitaxy (MBE)-grown crystals can be combined to understand the complex process of hydrogen and water adsorption on these rutile oxide surfaces. I will then move to the understanding of how alloying MBE-grown IrO2 with titanium can dramatically increase activity and how computational insight can help rationalize this surprising finding. Finally, I will describe our work on photocatalysis where single crystal semiconducting oxides (SrTiO3 and TiO2) are studied with ultra-fast spectroscopy to elucidate the surface structure and intermediates formed during light absorption. First principles computations are here used to understand the atomistic processes probed by the spectroscopy.

607b Combining Atomistic Simulation and Machine Learning to Establish High Throughput Yet Tunable Synthesis Strategies for Carbon-Based Nanomaterials

Rebecca Lindsey

Carbon nanoparticles (CNPs) are of tremendous interest for clean-energy technology due to the manifold of chemical, mechanical, electronic, and optoelectronic properties they can exhibit. However, practical application of these materials is hampered by current synthesis strategies. In particular, low-pressure techniques such as chemical vapor deposition and flame pyrolysis are well understood but relatively low throughput. Conversely, high-pressure methods like ultrasound cavitation and detonation can greatly enhance throughput (e.g., enabling rates of up to kgs/μs), but the underlying phenomena are not well understood due to the highly dynamic nature of these processes, hampering tunability.

In this presentation, we discuss recent collaborative efforts to bridge this gap through development of a new shockwave-based technique that enables rapid CNP synthesis and enhanced tunability over related high-pressure synthesis methods. A robust understanding of phenomena underlying material synthesis is critical for enabling tunability. Hence, we also discuss how we deploy ChIMES, our unique machine-learning-enhanced atomic-resolution simulation technique to simulate this synthesis process on scales overlapping with experiments. Ultimately, we show that our simulations provide a previously missing means of elucidating the complicated condensed-phase reaction driven phase separation and transformation processes that underly CNP shockwave synthesis.

607c Digital-Twin Battery Modeling and Simulations for Next-Generation Batteries

Yongmin Lee

Digital-twin-driven 3D modeling becomes as a pivotal tool in unraveling the intricacies of composite electrodes for next-generations batteries. Once we construct 3D structures with commendable precision, they become invaluable assets for scrutinizing structural, electrochemical, and mechanical dynamics under diverse conditions, from mechanical roll pressing to electrochemical testing. However, the initial challenge lies in crafting 3D structures with utmost accuracy, whether by reconstructing actual electrodes or creating virtual counterparts with limited data. Moreover, the choice of the optimal method is contingent upon the available time and computational resources.

In this endeavor, we embark on an exploration of how digital-twin 3D modeling reshapes the landscape of all-solid-state battery research. We delve into its applications, from visualizing the microcosm of composite electrodes to quantifying specific contact areas between electrode constituents and calculating effective electronic or ionic conductivity. Furthermore, we navigate the realm of simulations, unveiling their prowess in depicting voltage profiles, overpotentials, and lithium ion concentrations during cycling. Additionally, we unveil the successful simulation of mechanical behaviors within single electrode particles throughout the charging and discharging phases, marking a significant stride in our understanding.

Beyond the confines of electrochemistry, our inquiry extends to the mechanical realm, where we elucidate the structural deformation of composite electrodes through a comparative analysis employing finite element, finite volume, and discrete element methodologies. This comprehensive investigation not only sheds light on the multifaceted nature of all-solid-state batteries but also underscores the versatility and potency of digital-twin-driven 3D modeling in driving innovation within the realm of energy storage.

607d Accelerating Atomistic Simulation with Generative AI

Fei Zhou

The advent of generative artificial intelligence has provided new tools to model complex correlations in high-dimensional distributions. The recently proposed score dynamics (SD) is a general framework for learning accelerated evolution operators with large timesteps from molecular dynamics (MD) simulations.

SD enables us to take timesteps that are orders of magnitude larger than a typical MD time step. (1) In the position-only version of SD, we constructed graph neural network-based SD models of realistic molecular systems that are evolved with 10 ps timesteps. Both equilibrium predictions derived from the stationary distributions of the conditional probability and kinetic predictions for the transition rates and transition paths are in good agreement with MD tests on small molecules. (2) In the position-velocity version of SD, we have developed universally applicable acceleration techniques that allow one to take 10-100 fs timesteps in MD with little computational overhead, greatly extending the typical reach of molecular simulations.

607e Transformer-Based Hybrid Modeling and Control of Evolving, Nonlinear Processes

Joseph Kwon

Traditionally, the dynamic modeling of chemical processes has relied on first-principles models grounded in fundamental physics and chemistry laws. These models, primarily formulated through differential equations with constant parameters, enable the calculation of control actions optimizing process operations, taking both process and actuator limitations into account. However, the ever-evolving and nonlinear nature of chemical processes frequently calls for models with time-varying parameters. Motivated by these challenges, we have developed hybrid models that integrate system-agnostic first-principles dynamics with system-specific data-driven, time-varying parameters.

Our hybrid modeling framework incorporates a recent innovation: attention-based time-series transformers (TSTs) coupled with positional encoding. This marks a pioneering venture into applying the transformer algorithm – a cornerstone in ChatGPT's triumph - to nonlinear, time-varying processes. By analyzing data across both current and preceding time steps, the TST captures both immediate and historical changes in process states, granting a contextual insight on process dynamics, mirroring ChatGPT's textual context understanding. This TST-based hybrid model identifies correlations between process parameters and state variables. Its versatility is evident as it adapts to a spectrum of models - from density function theory to computational fluid dynamics - and scales, spanning from laboratory to extensive industrial environments. We will present applications of this hybrid modeling and control architecture, showcasing its utility from labs to industrial processes, made possible through partnerships with leading chemical process enterprises.

607f Active Learning of Multimetalic Single Atoms Catalysts

Byungchan Han

Single-atom catalysts (SACs) exhibit high activity for a wide range of sluggish reactions and allow performance tunability at atomic-level through the selection of central metals, ligand environments, and secondary metal sites. However, the design space with varying structures and compositions significantly hinders the fast and accurate identification of desired multimetallic SACs. In this work, we demonstrate a self-driving computational strategy for exploring binary metallic sites with varying combinations of 3d transition metals and different ligand environments, resulting in over 30,000 single atom sites for the electrochemical catalysis of the oxygen reduction and evolution reactions (ORR/OER). This approach is based on the density functional theory (DFT) calculations of binding energies and atomic descriptors as target properties and utilizes an equivariant graph neural network (GNN) as a surrogate model for predicting DFT labels directly from the atomic structure. The chemical environments learned by the GNN lead to capturing composition-structure-property relationships for the ORR/OER activity and selectivity. Active learning facilitates the investigation of the search space by balancing the exploration of unseen atomic structures with the exploitation of the active ones. The computational predictions of promising Co-Fe, Co-Co, and Co-Zn metal pairs are consistent with the state-of-the-art results of experimental measurements reported in the literature. The GNN-based analysis of multiple active sites on a single surface for target catalytic reaction can be extended to a broader class of multi-element high entropic materials systems.

607g Lattice Oxygen Kinetics in Nanostructured Ceria: Combining Graph Neural Network Multi-Scale Simulations and in-Situ DRIFT Characterization

Seokhyun Choung, Myeong Gon Jang, Jeong Woo Han

Ceria serves as a versatile catalyst in a wide range of reactions including automotive three-way catalysts and CO oxidation. The catalytic activity of ceria stems from its ability to readily release lattice oxygen, making ceria's reducibility crucial for understanding interactions with reaction intermediates on its surface. Density functional theory (DFT) studies have been conducted on well-controlled ceria surface systems at the atomic scale (~100 atoms). However, DFT calculations often rely on heuristic searches for active sites, overlooking geometric effects at the nanometer scale, limiting their applicability to recently proposed nanostructured ceria in the few-nanometer regime.

Here, we study the reduction process of lattice oxygen on the nanostructure-modified ceria systems via large-scale molecular dynamics using a pretrained Graph Neural Network (GNN). The pretrained GNN is successfully applied to the nanostructured ceria system, enabling simulations at larger time and size scales while maintaining DFT-level accuracy.

The lattice oxygen kinetics of the ceria surface is elucidated for the newly proposed 'Mace'-shaped ceria nanoparticles, where the morphology and size are precisely controlled in experiments during the hydrothermal ceria synthesis. Large-scale MD simulations reveal that facile lattice oxygen donation from ceria occurs under CO environment at the interface sites between rod (110) and cube (100) facets. Molecular dynamics simulations track the lattice oxygen donation process, akin to isotope tracing experiments, allowing the identification of the oxygen source. Furthermore, high-resolution characterizations using synchrotron and in-situ DRIFT techniques corroborate the facile oxygen donation at interfacial sites on ceria. By combining multi-scale simulations using a pretrained GNN and high-resolution experimental characterizations, this work provides a deep understanding of the oxygen kinetics in ceria and insights for rationally designing other metal oxide nanostructures.

545a Optimizing the Activity of Oxygen Reduction Reaction Via Structural Engineering of Support for Fe-N-C Electrocatalysts in Proton Exchange Membrane Fuel Cells

Jinwoo Lee

The scarcity and costliness of platinum group metals in PEMFC cathodes hinder commercialization. Despite the promise of Fe-N-Cs electrocatalysts as alternatives, their oxygen reduction reaction (ORR) performance falls short. To address this, two carbon support-structural engineering methods were devised. Firstly, defect adjustment in ZIF-NC carbon support, derived from zeolitic imidazolate frameworks, via CO_2 activation. Secondly, the introduction of mesopores into ZIF-NC using the block co-polymer soft-template method. Controlled degree of defect sites in supports enhance intrinsic activity of Fe-N_4 by modifying electronic structure such as charge distribution and spin state. Mesopores aid mass transport, improving active site accessibility. Additionally, the curved surfaces and carbon edge sites generated by mesopores alter the geometric structure of Fe-N_4, optimizing the adsorption energy of oxygen species. Support-structural engineered Fe-N-Cs electrocatalysts exhibit relatively outstanding ORR performance compared to pristine Fe-N-C. MEAs with these electrocatalysts exhibit impressive peak power densities, highlighting their high activity at the MEA level.

545b Enhancing Stability in Gas-Phase Reactions By Mitigating Metal Sintering Via Exsolution

Jeong Woo Han

Exsolution, a process whereby metal nanoparticles are liberated from ceramic hosts under specific reduction conditions, is a noteworthy phenomenon. These exsolved metal particles establish distinctive interfaces with perovskite hosts, either integrating into the surfaces or forming new chemical bonds. Not only do these nanoparticles offer abundant and uniformly dispersed catalytic active sites on oxide surfaces, but they also demonstrate robust thermal and redox stability.

Our ongoing investigation revolves around Pt exsolution catalysts supported by $LaFeO_3$ perovskite, showcasing exceptional endurance in demanding conditions, particularly in high-temperature water gas shift (HT-WGS) reactions. Structural analyses have unveiled significant preservation of particle dispersion upon exsolution, indicating an advantage in preventing metal sintering. To delve into the mechanisms enabling exsolution particles to maintain stability during reactions, we conducted diverse analyses such as in situ EXAFS, in situ DRIFT, density functional theory, and machine learning potential based molecular dynamics, focusing on factors like variations in intermediates, reaction pathways, and reactant sensitivity. Our ultimate aim is to pinpoint the reasons behind reduced coking tendencies.

These findings emphasize the remarkable stability of exsolution particles in HT-WGS environments and potential benefits in water or hydrocarbon-related reactions. Notably, their prospective integration into various catalytic applications holds substantial promise for addressing challenges such as metal particle agglomeration and coking.

545c Materials Interface Engineering for Membrane-Electrode-Assembly from Hydrogen Fuel Cell to Water Electrolyzer

Jin Young Kim

Advances in hydrogen fuel cell and water electrolyzer are urgently needed to increase the performance and reduce the cost of electrochemical conversion and utilization of hydrogen. Polymer membrane and catalyst electrodes are the core components to produce the electrochemical reaction and rapid progress in their development has increased its electrochemical-power conversion efficiencies and lifetime. Recent research and application activities include high performance noble metal alloy catalysts, thin film nanostructured electrocatalysts, durable catalyst support, polymer composite membrane, radical scavengers and oxygen-permeable ionomers. Moreover, the nanometer and micrometer scale interfaces in the membrane-electrode assembled layers are crucial for the scalable and sustainable electrochemical processes. In this talk, our recent results from these activities will be presented.

545d Laser-Synthesized Core-Shell ZIF-Based Nanocomposites As Efficient Electrocatalysts for Oxygen Reduction Reaction in Anion Exchange Membrane Fuel Cells

Mahshid Mokhtarnejad, Bamin Khomami

Synthesis of metal oxides (MO_x)/metal-organic frameworks (MOFs) into hybrid nanocomposites (HNCs), wherein transition metals combined with carbon (C) matrices, due to the excellent conductivity and high porosity present unique and cost-effective alternatives for oxygen reduction reaction (ORR) in anion exchange membrane fuel cells (AEMFCs). To that end, improving the performance of these materials by adjusting their composition, structure, size, and morphology through an efficient synthesis process is of great technological importance. Therefore, the Laser Ablation Synthesis in Solution in Tandem with Galvanic Replacement Reaction (LASiS-GRR) technique is used as a fast, facile, and environmentally benign approach for the synthesis and precise control of complex hierarchical bimetallic Zeolitic Imidazolate Framework (ZIF) structures. Specifically, this novel, one-pot, two-step LASiS-GRR process allows for the optimization of the composition, structure, size, and morphology of Co-based ZIF-67 encapsulated within Zn-based porous ZIF-8 crystals. The pyrolytic post-processing of these crystals leads to the development of HNCs exhibiting superior ORR electrocatalytic performance in AEMFCs. Our research demonstrates not only the impact of two distinct Co salt precursors on particle size and morphology during LASiS but also the exceptional performance and stability of post-pyrolytic (ZnO/ZIF@C) HNCs in AEMFC applications as non-precious-ORR-electrocatalysts compared to state-of-the-art Pt- and platinum-group metals (PGM)-based electrocatalysts.

545e Long-Life Reversible Li-CO_2 Batteries Based on Nanostructured Mn-Based Transition Metal Oxide Electrocatalysts

Jianda Wang

Combustion of fossil fuels to emit carbon dioxide (CO_2) becomes a major factor for green-house effect. Additionally, energy shortage presents a constraint on the progress of modern society. Many groups have provided different strategies for solving these two issues. However, most of the technologies still face the challenges of high-cost and low efficiency. Li-CO_2 batteries have emerged as one of the most promising energy storage devices by virtue of the advantages of cost savings, high theoretical capacity, and the excellent ability to capture green-house gas CO_2 via reaction with Li. However, high thermal stability Li_2CO_3, the main discharge product of Li-CO_2 batteries, hampers the improvement of cycle performance and practical capacity. In order to enhance cycle stability and capacity, it is necessary to find high efficiency electrocatalysts to help Li-CO_2 batteries decompose Li_2CO_3. In this work, we designed nanostructured AMn_2O_4-type (A= Co, Zn and Ni) transition metal oxide electrocatalysts though electrodeposition method, which can effectively improve cycle performance (> 200 cycles) and charge-discharge capacity (> 10000 mAh/g) of Li-CO_2 batteries. We performed spectroscopies and microscopies to gain a better understanding of the reaction mechanism of Li_2CO_3 formation and decomposition in Li-CO_2 batteries. This work not only paves a pathway to improve electrochemical performance of Li-CO_2 batteries, which would be beneficial for both energy storage and the reduction of CO_2 emissions, but also provides a novel strategy to design elaborately nanostructured transition metal oxide electrocatalysts.

545f From Powder to Large-Area Films - a Solution-Processable Route for Thin Films of Semiconducting 2D Tmds for Solar Energy Conversion

Rebekah Wells, Miao Zhang, Tzu-Heng (Bart) Chen, Victor Boureau, Charles Lhermitte, Marina Caretti, Hannah Johnson, Kevin Sivula

The remarkable semiconducting properties of 2D transition metal dichalcogenides (TMDs) are of great interest to researchers hoping to harness such traits for solar-to-chemical energy conversion (ex., H_2). Nano-TMDs not only provide a new range of thickness tunable optoelectronic properties, but also create a direct path towards ultrathin, flexible devices. However, scalable routes for large-area, high-quality TMD thin films remain a challenge for the commercialization of photoelectrochemical (PEC) devices[1].

We address this issue by developing methods for the solution-based fabrication of nanoflake TMD thin films for incorporation in large area solar energy conversion devices[2]. In this presentation we highlight a novel method for exfoliating commercially available TMD powders via electrochemical pellet intercalation[3]. Compared to traditional solution-processable methods such as ultrasonication, this method results in nanosheets with superior optoelectronic and photoelectrochemical properties. These specially engineered nanoflakes show higher photocurrent densities (J_{ph}), lower dark current densities (J_{dark}), and an absorbed photon-to-current efficiency (APCE) up to 90%. We examine the improved performance by comparing defect types and densities using fluorescent microscopy[4] and atomic resolution scanning transmission electron microscopy (AR STEM) paired with an integrated differential phase contrast (iDPC) technique[5].

To process our exfoliated materials into large-area thin films we report the further advancement of our preparation methods toward a continuous roll-to-roll deposition of TMD-based thin films from nanoflake dispersions using a liquid-liquid self-assembly technique[6]. We demonstrate reproducible printing of 100 mm wide TMD flake films on plastic substrates. Thus, we outline a pathway for fully solution-processable 2D-TMD devices beginning with as-received bulk TMD powder and finishing with large-area nano-flake thin films.

References

(1) Yu, X.; Sivula, K. Toward Large-Area Solar Energy Conversion with Semiconducting 2D Transition Metal Dichalcogenides. *ACS Energy Lett.* **2016**, *1* (1), 315–322. https://doi.org/10.1021/acsenergylett.6b00114.

(2) Wells, R. A.; Sivula, K. Assembling a Photoactive 2D Puzzle: From Bulk Powder to Large-Area Films of Semiconducting Transition-Metal Dichalcogenide Nanosheets. *Acc. Mater. Res.* **2023**, *4* (4), 348–358. https://doi.org/10.1021/accountsmr.2c00209.

(3) Wells, R. A.; Zhang, M.; Chen, T.-H.; Boureau, V.; Caretti, M.; Liu, Y.; Yum, J.-H.; Johnson, H.; Kinge, S.; Radenovic, A.; Sivula, K. High Performance Semiconducting Nanosheets *via* a Scalable Powder-Based Electrochemical Exfoliation Technique. *ACS Nano* **2022**, *16* (4), 5719–5730. https://doi.org/10.1021/acsnano.1c10739.

(4) Zhang, M.; Lihter, M.; Chen, T.-H.; Macha, M.; Rayabharam, A.; Banjac, K.; Zhao, Y.; Wang, Z.; Zhang, J.; Comtet, J.; Aluru, N. R.; Lingenfelder, M.; Kis, A.; Radenovic, A. Super-Resolved Optical Mapping of Reactive Sulfur-Vacancies in Two-Dimensional Transition Metal Dichalcogenides. *ACS Nano* **2021**, *15* (4), 7168–7178. https://doi.org/10.1021/acsnano.1c00373.

(5) Yücelen, E.; Lazić, I.; Bosch, E. G. T. Phase Contrast Scanning Transmission Electron Microscopy Imaging of Light and Heavy Atoms at the Limit of Contrast and Resolution. *Sci Rep* **2018**, *8* (1), 2676. https://doi.org/10.1038/s41598-018-20377-2.

(6) Wells, R. A.; Johnson, H.; Lhermitte, C. R.; Kinge, S.; Sivula, K. Roll-to-Roll Deposition of Semiconducting 2D Nanoflake Films of Transition Metal Dichalcogenides for Optoelectronic Applications. *ACS Appl. Nano Mater.* **2019**, *2* (12), 7705–7712. https://doi.org/10.1021/acsanm.9b01774.

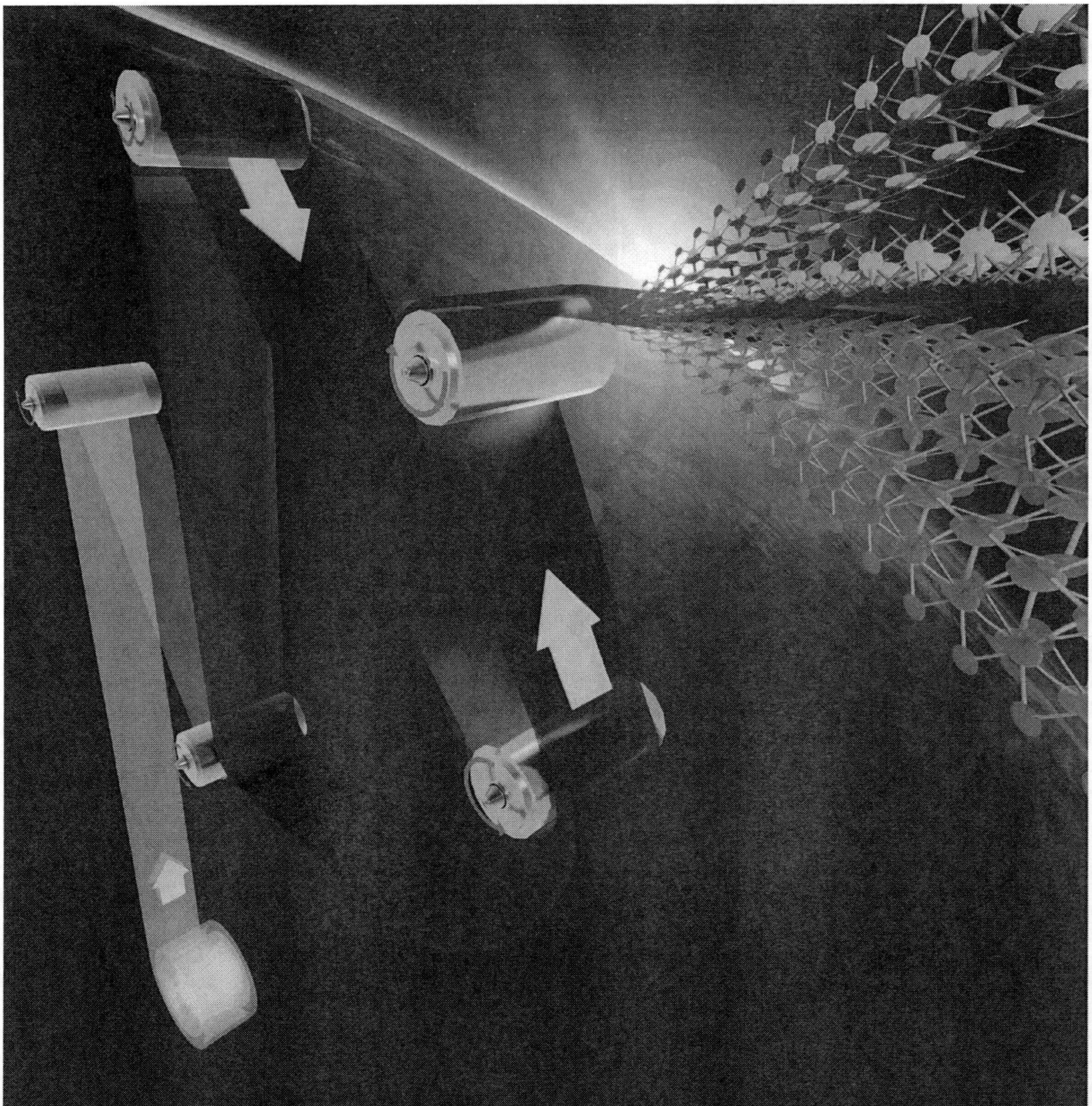

545g Investigating the Effects of Photothermal Nanoparticles on Diel's Alder Reaction Towards Development of Light Induced Recycling of Thermosets.

Arnob Dipta Saha, Julio Cubillas, Youngmin Lee, John McCoy, Sanchari Chowdhury

Thermosets are exceptionally ideal for a wide variety of applications, including adhesives, electronics, automotives, aerospace, and household gadgets, because of their remarkable mechanical qualities, strong chemical resistance, and excellent thermal stability However, the recycling of thermosets has posed a significant challenge due to their irreversible crosslinked structure limiting their reprocessability and recyclability. Furthermore, the absence of precise heating in specific areas and the low thermal conductivity of epoxies can render the recycling process utilizing bulk heating ineffective. This study involves the incorporation of refractory plasmonic Titanium Nitride (TiN) and photothermal Carbon Black (CB) nanoparticles into epoxy. These nanoparticles can absorb light and generate localized nanoscale and macroscale heat, respectively. This heat can then drive reversible reactions, such as Diels-Alder reactions, which enable the recycling of the epoxy material. Nanoparticles loading and their dispersion in the polymer matrix is optimized to maximize their photothermal efficiency. We used in situ FTIR to study the kinetics of both forward and reverse Diels Alder reactions driven by photothermal effects and compared that with conventional heat driven reaction kinetics. Moreover, the study examined the kinetics in the presence of heat using both CB and TiN and compared it to the neat Diels-Alder epoxy resin. One of our major observations is that, despite the sample having a similar bulk temperature, the forward Diels-Alder reaction kinetics with light-induced photothermal heat generation varies dramatically from conventional heating. To the contrary, the kinetics of the reverse Diels-Alder process was not significantly altered by the presence of nanoparticles and different stimuli. It implies that the Diels-Alder reactions may be impacted differently by reactant interaction with nanoparticles and local heat generation surrounding the nanoparticles. Further research is being conducted currently to understand the underlying mechanisms of photothermally induced Diels-Alder reaction.

545h Optimizing the Electrophoretic Assembly of Silica Nanoparticles: The Role of NaCl Electrolyte and Process Parameters

Srabani Karmakar, Swomitra Mohanty

The emergence of ordering in silica nanoparticle monolayers through electrophoretic deposition (EPD) offers a promising avenue for achieving controlled assembly in the synthesis of nanomaterials. This work investigates the process of organizing nanoparticles during EPD to understand the fundamental processes that govern ordering phenomena. With a 0.01M NaCl aqueous electrolyte solution serving as the deposition medium, we analyze the effects of essential factors on the deposition rate and final structure, including particle concentration, deposition duration, pH, electrolyte conductivity, and current density. Through TEM analysis and experimental characterization, we demonstrate the importance of the 0.01M NaCl aqueous electrolyte in facilitating efficient particle migration and deposition, particularly at optimal silica nanoparticle concentrations. The precise concentration of NaCl in the electrolyte ensures a homogeneous suspension. It inhibits nanoparticle agglomeration, which is essential for uniform deposition and the development of well-ordered structures. Furthermore, we discover that the deposition rate and the ensuing ordering of nanoparticles are highly susceptible to changes in particle concentration, deposition duration, and current density. The knowledge gained from this study contributes to our understanding of the colloidal assembly of silica nanoparticles and its applications in energy, sensing, photonics, and other fields. It also has practical implications for the customized fabrication of functional nanomaterials with improved properties.

211a Multiscale Modeling Approach to Solid State Polymer Electrolyte for Li-Ion Battery

Seung Soon Jang

Solid-state polymer electrolytes emerge as a promising alternative to liquid electrolytes, addressing well-documented challenges by offering improved safety, thermal stability, and resistance to dendrite growth, along with flexible integration into battery designs. Despite their advantages, several critical obstacles remain, including low ion conductivity, constrained operational temperature ranges, and electrode material compatibility. To overcome these, a systematic understanding of their molecular structures, thermal properties, and ion transport mechanisms is crucial. This study leverages multiscale modeling and simulation approach to investigate carbonate-based solid polymer electrolytes (SPE) with various side chain lengths and small molecular lubricants, in comparison to the traditional liquid carbonate electrolytes. Our MD simulations focus on analyzing the impact of polymer chain design on the nano-segregated structure and thermal behavior. By contrasting these with the properties of liquid electrolytes like dimethyl carbonate and ethylene carbonate, we aim to offer a detailed view of the nanostructure, thermal dynamics, and lithium-ion transport in polymer electrolytes. Our research seeks to provide insights into engineering polymer materials with optimized nanostructures and enhanced lithium-ion transport, contributing to the development of safer and more efficient battery technologies.

211b Constructing Reliable Lithium Metal Surface with Functionalized 2D Mxene Mono-Layer As Coating

Junjie Niu

In order to achieve high-energy-density lithium-ion batteries to meet the rapidly growing market in electrical vehicles (EVs) and portable electronic devices, Li metal is considered as the most promising candidate as anode. However, the dendrite forming, low Coulombic efficiency, and unstable solid electrolyte interphase (SEI) pose big challenges in applying Li metal in batteries. To date several strategies on stabilizing the Li metal electrode have been attempted including but not limited to 1) add electrolyte additives and increase lithium salt concentrations, 2) use solid electrolytes with high ionic conductivity to mechanically inhibit the dendrite growth, 3) develop electrochemically stable artificial SEI, and 4) pre-store Li in stable hosts. In this presentation, I will talk about our recent progress in stabilizing the Li metal surface using two-dimensional MXene coating as anode in lithium ion batteries. First we will introduce covalent organic framework (COF) modified MXene mono-layer on Li metal. The lithiophilic coating promotes the uniform Li nucleation by homogenizing the Li ion flow and decreasing the nucleation barrier, resulting in a dendrite-free lithium metal anode. Second a dual-layer interphase that consists of an *in-situ* formed lithium carboxylate organic layer and ultra-thin BF_3-doped monolayer MXene was formed on Li metal surface. The honeycomb-structured organic layer increases the wetting of electrolyte due to the large surface area, leading to a thinner SEI. While the BF_3-doped monolayer MXene provides abundant active sites for lithium homogeneous nucleation and growth, resulting in about 50% reduced thickness of inorganic-rich components among the SEI layer.

211c Advancing Lithium Battery Safety and Performance: Aqueous Electrolyte Design with Ternary Eutectic System

Youngbi Kim, Jeong Woo Han

Lithium-ion batteries are widely used in modern electronics and portable power storage systems; however, safety concerns remain a significant consideration. Particularly, the risk of fire and explosion associated with non-aqueous electrolytes underscores the ongoing need for research and innovation. One solution to these issues involves the use of aqueous electrolytes. The recent concept of 'water-in-salt' (WISE) electrolytes has emerged, wherein the presence of numerous salt aggregates within the electrolyte prioritizes salt decomposition over solvent decomposition, forming robust interphase layers on electrodes and enhancing battery system stability. In the realm of aqueous electrolytes, efforts have been made to further expand the electrochemical stability window of WISE by reducing the number of water molecules in the lithium ion solvation shell. In this study, acetamide was introduced as a third component. Ternary eutectic electrolytes were formed by altering the ratio of lithium salt, acetamide, and water molecules, with the 2:3:1 ratio showing the most significant extension of the electrochemical stability window. Comparisons among the systems were observed through Molecular Dynamics (MD) and Density Functional Theory (DFT) analysis. In order to validate the micro-scopic bonding network, we computed the radial distribution functions (RDFs), and the results showed that the Li-TFSI interaction peaked at 2:3:1, with subsequent values occurring at 2:4:1 and 1:3:1. Li-TFSI-Li-TFSI-Li was also measured, and the distance between them was closest at 2:3:1. This demonstrated that the aggregation at 2:3:1 had a large number of solvated structures. The closest distance of 2.625Å of Li-H_2O in 2:3:1 indicates that 2:3:1 has more compact solvent shell than the other two compositions, although having the smallest RDF pick. The most solvated structures from each system were chosen based on the MD results. Through the use of DFT calculations, it was determined that the OH binding of H_2O strengthened in the 2:3:1 solvent structure, implying a weakening of the H_2O network. Electronic structure changes are shown by the projected density of states as a result of solvation structure modifications. The lowest unoccupied molecular orbital (LUMO) value of H_2O rises upward relative to the other two systems, the HOMO-LUMO gap of H_2O widens, and the HOMO-LUMO of TFSI increases. As the gap changes, H_2O breakdown is inhibited and the formation of a high-quality passive layer on the surface is facilitated. Moreover, an electrode/electrolyte interphase is produced during TFSI breakdown, which enhances electrochemical stability. Lastly, we computed the situation of raising the interfacial layer's thickness because water breakdown can still happen because of electron tunneling, even when the potential window is raised. As a result, DFT determined that as SEI thickness increases, the tunneling coefficient reduces, and MD verified that water decomposition decreases. This study provides this novel electrolyte design that greatly expands the electrochemical stability window while retaining the high safety of aqueous electrolytes by combining non-flammable asymmetric donor-acceptor molecules with aqueous solutions. This breakthrough goes beyond lithium-ion technology and advances the creation of high-energy-density, safe, and affordable aqueous batteries.

211d Efficient Strategies to Accelerate Anodic Reactions in Electrolysis Systems for Hydrogen Production

Hongmei Luo, Meng Zhou

Abstract

Electrochemical water splitting is an economical and green approach to producing clean hydrogen, a promising technology to alleviate the energy crisis. A high operating cell voltage is needed to achieve the desired current density due to the sluggish kinetics of oxygen evolution reaction (OER). Herein, we demonstrated two strategies to improve the OER activity of the cobalt-nickel sulfides on Ni foam (Co-S/Ni3S2@NF): adding glycerol and increasing the temperature. A current density of 20 mA cm^{-2} could be reached at an extremely low potential of 1.35 V (vs. RHE) by adding 0.1 M glycerol to the alkaline electrolyte, indicating glycerol oxidation reaction (GOR) is a possible anodic reaction to replace the sluggish OER. The estimated activation energy of Co-S/Ni3S2@NF for OER is higher than that for GOR, indicating the OER process with high kinetic energy barriers. By adding glycerol and increasing the temperature, the anodic current density at ~1.6 V was observed to increase tenfold, providing solutions to improve electrolysis efficiency. These findings provide valuable clues to overcoming the sluggish kinetics of the anodic reaction process and improving electrolysis efficiency.

211f Laser-Synthesized Zn/Co-Based Metal-Organic Frameworks Composited with Reduced Graphene Oxide As High-Performance All-Printed Energy Storage Devices

Mahshid Mokhtarnejad, Bamin Khomami

Supercapacitors (SCs) with rapid recharge rates and substantial power output bridge the gap between conventional capacitors and batteries, making them essential components in energy storage applications. Metal-organic frameworks (MOFs) have emerged as promising candidates for SC technology due to their unique properties, including high porosity, tunable pore size distribution, and exceptional structural adaptability. When composited with other materials, such as carbon-based compounds, graphene, and graphene oxide, to produce hybrid nanocomposites (HNCs), they achieve high conductivity and enhanced chemical stability, making them ideal as high-performance electrode materials in SCs. This study presents a rapid and cost-effective method for synthesizing bimetallic MOFs composited with reduced graphene oxide (rGO) for energy storage applications. Our results demonstrate that Laser-derived bi-MOF-rGO HNCs exhibit high specific capacitance and power density, showcasing their potential for SC technology. Furthermore, we successfully integrate these HNCs into an all-printed active electrode material layer using sequential inkjet printing, offering a pathway for developing high-performance printed electronics.

211g Hollow ZnSe/$SnSe_2$ Cubes with Urchin Spines-like Rods Incorporated Carbon Nanofibers As Anodes for Sodium-Ion Batteries

Zhi Liu, Xing Li, Zhen Wei, Kenneth Gordon, Cameron Romero, Ling Fei

The exploration of Sn-based anode materials has gained considerable momentum owing to their remarkable theoretical capacity and energy density. Among them, alloy-based materials provide a promising route for improved stability and enhanced reaction kinetics during a charging/discharging process. In addition, combining metal compounds with carbon matrix represents a common and effective strategy to further enhance structural stability and integrity for practical application of batteries. Herein, a unique three-dimensional (3D) carbon nanofibers (CNFs) scaffold comprising of hollow ZnSe/$SnSe_2$ cubes with urchin-like spines was rationally designed by hydrothermal, electrospinning and carbonization processes, as a free-standing anode, resulting in superior electrochemical performance. Selenides are preferentially chosen over other chalcogenides, attributable to their balanced characteristics in electrical conductivity for electron transfer, interlayer spacing for Na ion transfer, and theoretical capacity. Hollow cavity in ZnSe/$SnSe_2$ cubes increases specific surface area, exposes more accessing sites in electrolyte for reactions, and compensates volume expansion. Notably, heterostructures of ZnSe/$SnSe_2$ can facilitate fast ion/electron transfer and kinetics of reactions in batteries, through built-in electric field in the formed interfaces between ZnSe and $SnSe_2$ nanoparticles. To ensure the overall integrity of anode, carbon nanofiber matrices serve as another reliable protective shells and excellent conductive skeletons, preventing selenides from direct contact with electrolyte and enhancing conductivity of whole structure. This innovative approach combines the merits of various components for great integrality of electrodes and ultimately broadens the routes for developing high-performance Sn-based anode materials in SIBs.

211h Thermal Reliability and Degradation of $CsPbBr_3$ Nanocrystals and Their Impact on Optical Properties and Structural Integrity in Optoelectronic Applications

Moonchul Park, Suchi Guha, Yangchuan Xing, Jaewon Lee

The rapid advancement in nanotechnology has brought cesium lead bromide (CsPbBr3) nanocrystals (CPB NCs) to the forefront of materials for optoelectronic applications. As a part of the all-inorganic perovskite family, CPB NCs have attracted significant attention due to their remarkable optical properties, such as a direct bandgap and high photoluminescence quantum yield, making them highly desirable for use in light-emitting diodes (LEDs), solar cells, and photodetectors. The thermal reliability of CsPbBr3-based devices is a crucial factor for their performance and stability, as it affects their ability to withstand thermal stress during operation or fabrication, impacting both their optical properties and structural integrity. However, like other perovskite materials, CPB NCs are susceptible to thermal degradation, which can manifest as phase transitions, changes in crystal structure, and alterations in optical properties. In this paper, we discuss the thermal reliability of CPB NCs regarding their changes in optical properties and thermal degradation regarding at various temperatures. CPB NCs synthesized by hot injection method dispersed in n-heptane as colloidal were stored at -20°C, 25°C and 70°C while TEM images, Photoluminescence (PL) intensity, UV-vis absorbance and X-ray diffraction were conducted to observe thermal stability of CPB NCs. The particle size of CPB NCs stored at 70°C increased by 46.8 % over 30 days, but the size of CPB NCs stored at -20°C and 25°C increased by only 7.83 % and 13.4 % respectively. PL intensity peak for CPB NCs stored at 70 °C showed red-shift of 5nm over 30 days, while CPB NCs stored at - 20°C and 25°C showed no shift. XRD patterns showed peaks corresponding to the (100), (110), and (200) planes of CPB NCs. At -20°C and 25°C, the intensity ratio of the (100) to (110) peaks remained relatively unchanged over 30 days, while the intensity of the (110) peak increased relative to the (100) at 70 °C after 10 days. Additionally, storage at 70°C led to shifts in the (200) XRD peak, indicating crystal structure distortion by analyzing FFT of TEM images. The structures of CPB NCs stored at 25 °C remained unaffected under TEM beam exposure, while CPB NCs stored at 70 °C led to crystal degradation under TEM beam with a dose rate of 5780 $e/Å^2s$ for 300 seconds.

82a NSEF Young Investigator Award Winner: Light Emission and Lasing in Organic Semiconductor-incorporated Perovskite (OSiP)

Letian Dou

Halide perovskites are exciting new semiconductors that show a great promise in low cost and high-performance optoelectronics devices. However, the poor stability is limiting their practical use. In this talk, I will present a molecular approach to the synthesis of a new family of hybrid material – Organic Semiconductor-incorporated Perovskite (OSiP), which are more versatile and intrinsically stable. Energy transfer and charge transfer between adjacent organic and inorganic layers are extremely fast and efficient, owing to the atomically flat interface and short interlayer distance. In addition, the rigid conjugated ligand design dramatically enhances their chemical stability, suppresses solid-state ion diffusion, and modulates electron-phonon coupling, making them useful in many applications, particularly solid-state lighting. Using these stable hybrid materials, we demonstrate efficient light emission and amplification in single crystalline nanostructures, epitaxial heterostructures, and polycrystalline thin films.

82b Nanoscale Science and Engineering Forum Award Winner: Quantum Dots for Clinical Diagnostics: A Commercialization Story

Jessica Winter

Quantum dots are semiconductor nanoparticles that were first discovered in the early 1980s. Because of their small size, they experience quantum confinement, which gives them many unique properties important for fluorescence imaging, including broad excitation spectra, narrow emission bandwidths, and size tunable fluorescence. However, despite initial attempts in the late 1990s and early 2000s to introduce them for biological imaging, they failed to replace traditional fluorophore dyes. My research for the last 25 years has focused on uncovering and solving critical roadblocks to the application of quantum dots in clinical biological imaging. Quantum dot properties and typical formulations for biological imaging will be discussed. Among the problems plaguing these formulations is the need to transfer nanoparticles synthesized in the organic phase to aqueous media. We showed that the challenges encountered during this process, including aggregation and loss of fluorescence, result primarily from interactions with aqueous media that can lead to ligand loss and surface oxidation. To address these challenges, we developed new nanomanufacturing technologies that encapsulate quantum dots in polymer micelles. These technologies preserve original ligand structures and minimize exposure of the nanoparticles to aqueous media to maximize retention of quantum dot properties. These technologies have also enabled scalable production of these materials, leading to their commercialization. In 2012, I founded Core Quantum Technologies (CQT), a company dedicated to producing quantum dots for clinical diagnostics of hematological cancers. CQT has faced significant obstacles, including the funding "valley of death", personnel issues, and Covid. The difficulties of hands-on participation in a start-up and navigating to a post-revenue company will be described. CQT now offers a vibrant product line of quantum dot and magnetic quantum dot reagents, an example of a successful the "bench to bedside" transition. These nanomaterials improve upon the accuracy and precision of current diagnostic technologies and offer new avenues for personalized medicine and patient therapeutic design.

142a Graph Theoretic Characterisation of Chirality in Complex Nanodendrimers

Alain Kadar, Vera Kuznetsova, Anita Gaenko, Engin Er, Tao Ma, Kody Whisnant, Bing Ni, Ji-Young Kim, Yurii K. Gun'ko, Nicholas A. Kotov

Organic, polymeric, and inorganic materials with branching segments radially diverging from the central core are known for unique biological, biomedical, rheological, and other properties. They are referred to as dendrites, dendrimers, arborols, and fractal particles with majority of them displaying stochastically bifurcating structures. There is no established pathway to relate their complex structure combining both order and disorder to measurable macroscale properties. Here, we investigate nanostructured dendrimer-like gold particles with various branching generations Unlike previous nanostructure assemblies with star-, flower-, urchin- and hedgehog-like morphologies, the nanodendritic particles in this study are flat with spikes predominantly confined to one plane. Additionally, they are chiral with multiscale mirror asymmetry determined by *L/D*-enantiomers of cysteine used during synthesis. Their branching segments can have preferentially left- or right-handed twist leading to polarization rotation. The complex structure of the particles can be quantified using graph theory (GT) which is capable of accurately describing the combined order and disorder in their architecture. Here, we show how classical GT parameters are augmented with geometry to describe this complex structure. Furthermore, we show how electrodynamic simulations may be used to inform the definition of a new parameter, the spike branch factor (SBF) that accurately captures the effect of structure on the optical asymmetry factor (i.e. g-factor). We expect that these parameters will provide a pathway for property-centric design of the complex particles.

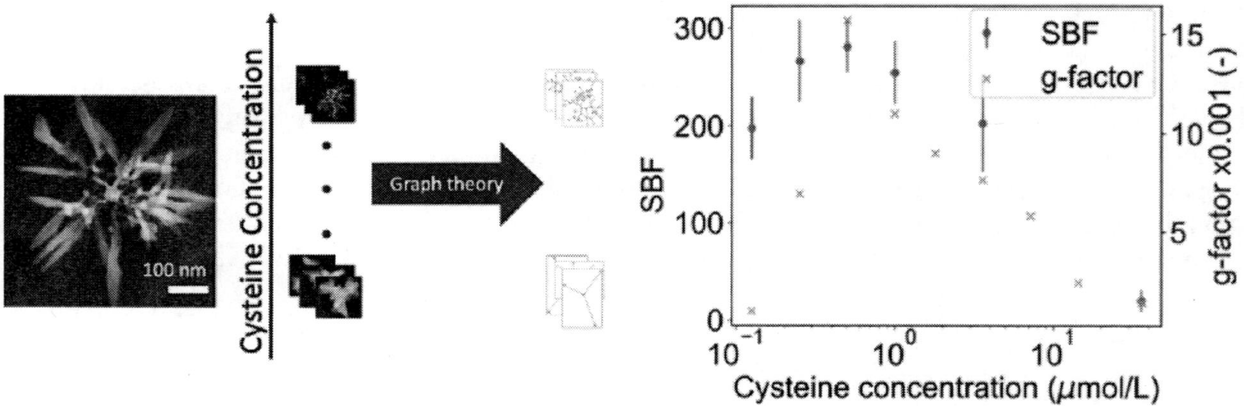

142b Tailoring Dissolution for Phothermal Patterning of Semiconducting Polymers

Joaquin Mogollon Santiana

The industrial advancement of semiconducting polymers (SPs) encounters a significant challenge due to the absence of an inexpensive, rapid, and viable patterning technology capable of producing sub-micron features. Our study dives into photothermal patterning (PPL) and its potential to address this challenge by enabling the precise tuning of the temperature at which a polymer dissolves, facilitating pattern formation. PPL operates on the principle of polymer dissolution: a thin film is submerged in a binary solvent mixture containing both a good and a bad solvent for the polymer at a specific ratio which can be tuned for instance 1-2, Dichlorobenzene (DCB) and Cyclohexanone (CHN). At room temperature, the polymer remains insoluble; however, upon exposure to a laser, the polymer undergoes a non-emissive process, heating up the surrounding solvent and inducing dissolution of the film, thus creating patterns. The selection of the solvent mixture ratio allows for the customization of the dissolution temperature of the polymer. In this study, we experimentally explored the dissolution temperature of poly(3-hexylthiophene) (P3HT) using various combinations of solvents, demonstrating the tunable control over dissolution. Furthermore, we investigated how controlling the molecular weight and polydispersity index (PDI) of the polymer impacts both the dissolution range and the quality of the patterns produced. Overall, our method, coupled with the associated theoretical model, lays the groundwork for the development of a cost-effective and rapid photo-patterning technology for SPs. This advancement opens up new avenues for industrial applications in the micro-fabrication of organic electronic devices.

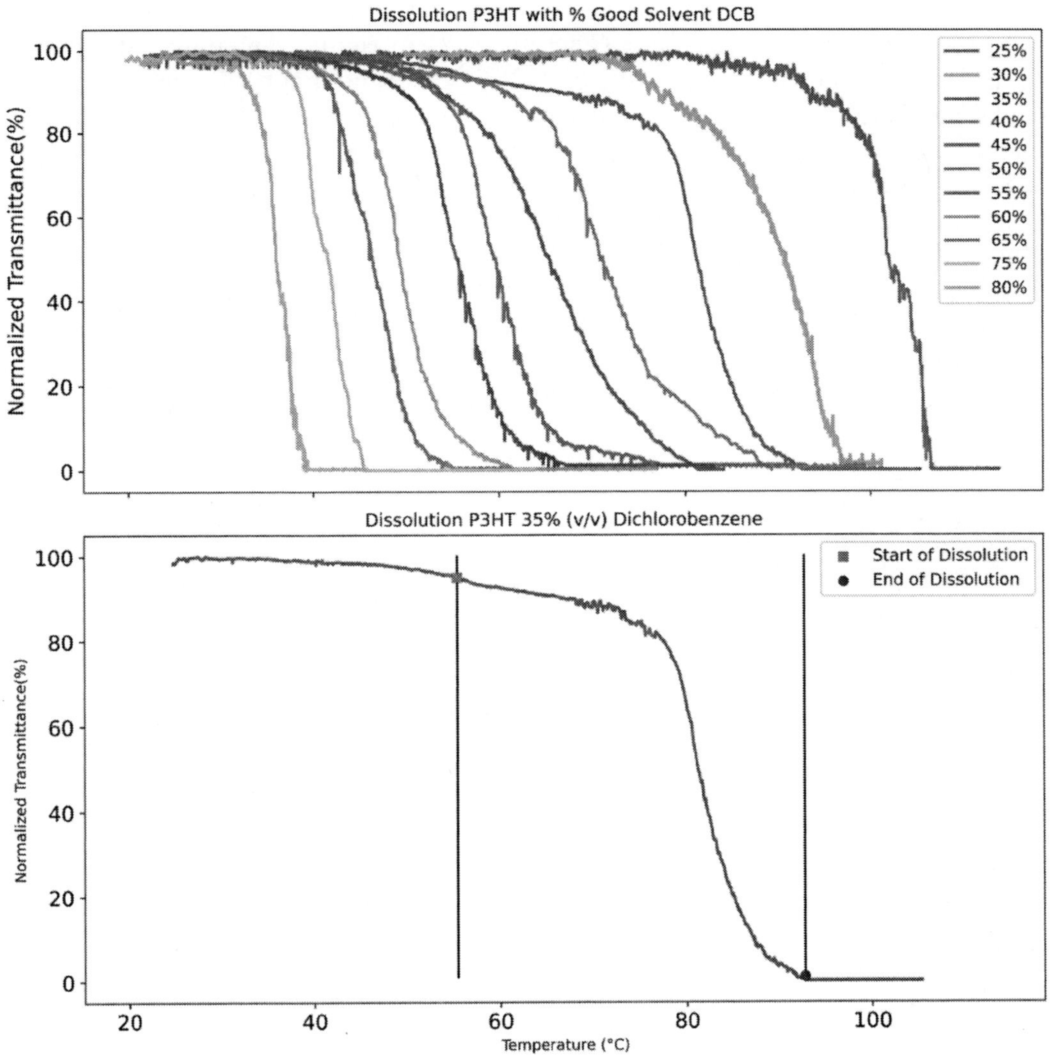

142c Direct Patterning of Plasmonic Nanohelicoids Via Circularly Polarized Light

Ji-Young Kim, Connor McGlothin, Emine Sumeyra Turali-Emre, Minjeong Cha, Wonjin Choi, Zechariah Pfaffenberger, Julie Biteen, Nicholas A. Kotov

Chiral plasmonic surfaces, characterized by nanohelicoidal structures forming 'forests,' exhibit substantial optical rotation but typically involve complex and expensive multi-step nanofabrication processes, limiting compatibility with diverse substrates. Additionally, achieving large-scale photonic patterns on polymeric and flexible substrates remains a challenging task. In this study, we present a breakthrough in substrate-tolerant direct-write patterning, demonstrating the fabrication of silver nanohelicoids with locally variable optical activity using circularly polarized light (CPL). Centimeter-scale chiral plasmonic surfaces can be produced within minutes using inexpensive medium-power lasers. The growth of nanohelicoids is governed by the symmetry-broken site-selective deposition and self-assembly of silver nanoparticles (NPs). The ellipticity and wavelength of incident photons serve as parameters to dynamically control the handedness and size of the printed silver nanohelicoids, allowing on-the-fly modulation of nanohelicoid chirality during direct writing. This process offers straightforward pathways to create complex multifunctional metasurfaces with high polarization rotation and fine spatial resolution. Our computer-driven direct-write system extends its capabilities by printing local patterns with various optical activities over distances spanning four orders of magnitude. Importantly, direct-write printing by CPL involves no organic ligand materials, promising in chiroptical diagnostics by enhancing sensitivity to the dielectric environment and eliminating the charge transport dilemma of solution-processed nanomaterials. We conducted spectroscopic detection of chiral analytes based on the CD peak wavelength shift of the printed chiral silver pattern. CD spectra of the pattern exposed to pepsin gradually red-shift with increasing concentration, demonstrating a limit of detection below 100 pM. Enantiomer selectivity is also confirmed through different absolute wavelength shifts observed upon the absorption of L- and D-lysine.
Processing simplicity, high polarization rotation, and fine spatial resolution of the light-driven printing of stand-up helicoids provide a rapid pathway to chiral plasmonic surfaces, accelerating the development of chiral photonics for health and information technologies.

142d Enhancing Quantum Dot Light Emitting Devices: Sub-Bandgap Turn-on and Electrobrightening Via Advanced Charge Transport Layer Engineering

Doh C. Lee

Over the past decade, QD-LEDs have experienced remarkable improvements in device efficiency, yet the mechanism behind barrierless hole injection at voltages below the bandgap remained elusive. Our research team has identified the Fermi level alignment, influenced by surface states, as the key factor enabling this phenomenon. This alignment, coupled with the macroscopic electrostatic potential gain from the restructured energy environment, facilitates hole injection into quantum dots. Contrary to previous assumptions, we demonstrate that the energy level alignment significantly outweighs the local carrier injection barrier alteration caused by the Coulombic force of a charge in quantum dots. Our findings elucidate the mechanism of barrierless carrier injection in QD-LEDs and propose a broader design principle for efficient electroluminescent systems utilizing nanocrystal emitters. Additionally, I will discuss the electrobrightening effect observed in quantum dots using ZnMgO as an electron transport layer instead of ZnO.

142e Nano-Scale Phenomena Control Growth and Stability of CO2 Hydrates

Alberto Striolo

Clathrate hydrates are solid inclusion compounds that promise to be extremely useful both for the production of natural gas and for the sequestration of carbon dioxide. The technoeconomic feasibility of both these processes, as well as others that inlcude separations and water purification, depends strongly on fundamental principles such as stability, transport and kinetics rekated to the formaton of the hydrates. Many experimental investigations are conducted, revealing interesting and sometimes puzzling properties of these compounds. To obtain a deep understadning of these systems it is necessary to develop a theoretical framework inclusive of these observations. As a first step, our group has used molecular simulations for uncovering some of the nano-scale phenomena that are likely responsible for the macroscopic observations. We will discuss here how interfacial phenomena occurring at the interface between solid and fluid interfaces dictate some of the macroscopic observables. Some of our recent insights will be presented here, towards identifying experimental processes that could be used for assessing their reliability.

142f Kinetics of Growth and Reaction Mechanism Monitoring of the Synthesis of Polydopamine Nanoparticles Using DLS and Via UV-Vis-NIR Spectrophotometry.

Carlos Cuervo, Cesar Luis Peña Sterling, Andres David Salinas Quinchanegua, Milena Vega, Julio C. Vargas

In recent years, the field of materials science and engineering has witnessed a growing interest in nanoscale materials with unique physical, chemical, and mechanical properties. These materials, derived from easily accessible, cost-effective, and renewable precursors, hold enormous potential for applications in surface coatings, biomedicine, environmental catalysis, pharmaceuticals, and renewable energy [Wu M., 2020].

This study investigates the kinetic behavior of dopamine (DA) polymerization through auto-oxidation in the presence of air to produce polydopamine (PDA) nanoparticles. The synthesis of PDA nanoparticles is crucial due to their potential applications in various fields. Specifically, PDA nanoparticles have been explored to remediate contaminated water, develop pharmaceutical products, and effectively treat diseases such as cancer. Additionally, they offer solutions for sustainable energy storage. Since nanoparticle properties exhibit distinct characteristics based on their size, this study focuses on examining the influence of reaction temperature and alkali concentration on the average diameter of PDA nanoparticles. [Djermane, R., 2021] Based on experimental data, a model is proposed to describe the variation in nanoparticle diameter as a function of temperature and alkali concentration, considering that the Avrami model proposed as an initial model does not show a direct relationship between the kinetic constant and temperature.

Furthermore, it is suggested to utilize near-infrared spectroscopy (NIR) and UV-Vis technique as tools for monitoring and analyzing this reaction. These techniques aid in comprehending the physicochemical properties of the reaction by detecting the formation and consumption of specific chemical groups throughout the process, identifying absorption peaks or characteristic bands in the spectrum [Pegueros Gutierrez A., 2010]. Specifically, UV-Vis allows determining the concentration of the compounds involved, with results presented in Figure 1, while NIR, with results such as those shown in Figure 2 and Figure 3, is employed to characterize and identify the properties of the compounds by recognizing specific peaks in the spectrum.

On the other hand, considering that current polydopamine formation reactions are conducted in batch processes, it is acknowledged that this mechanism entails operational disadvantages such as lower production capacity, in addition to potential challenges in maintaining consistent product quality and increasing production costs. Therefore, understanding the reaction kinetics paves the way for scalability from a batch to a continuous process, which is a crucial aspect with significant industrial implications. Given the importance of the size of polydopamine nanoparticles in obtaining properties with antineoplastic behavior, and their potential use as therapy in cancer, comprehending the influence of operating conditions on the properties of the nanoparticles is crucial for the implementation of continuous production processes [Nieto C., 2021].

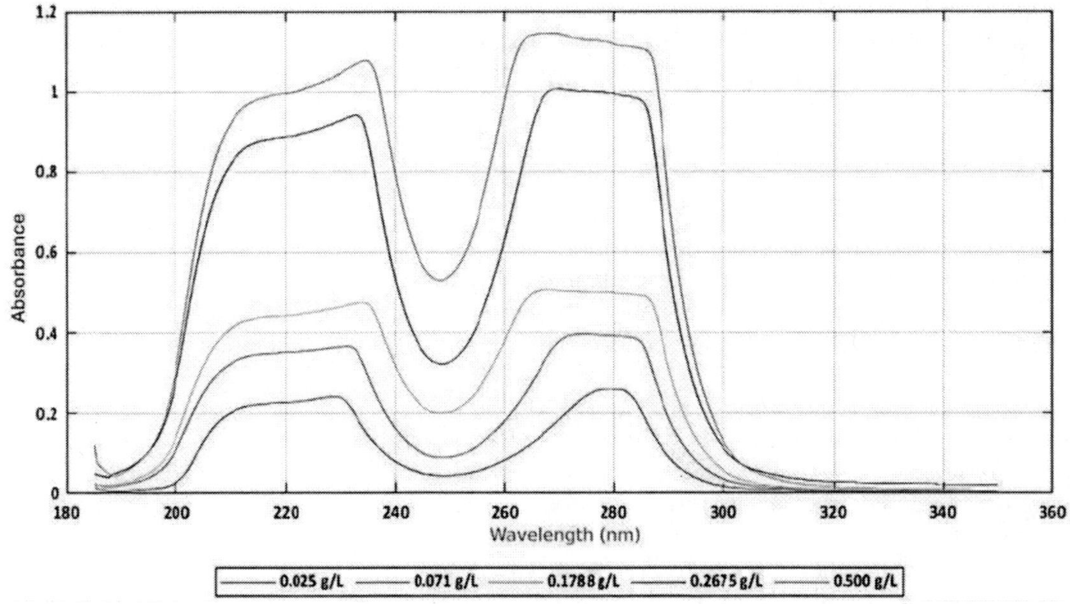

Figure 1. *UV-Vis spectrum scanning for 5 different dilutions of dopamine (DA) at varying concentrations.*

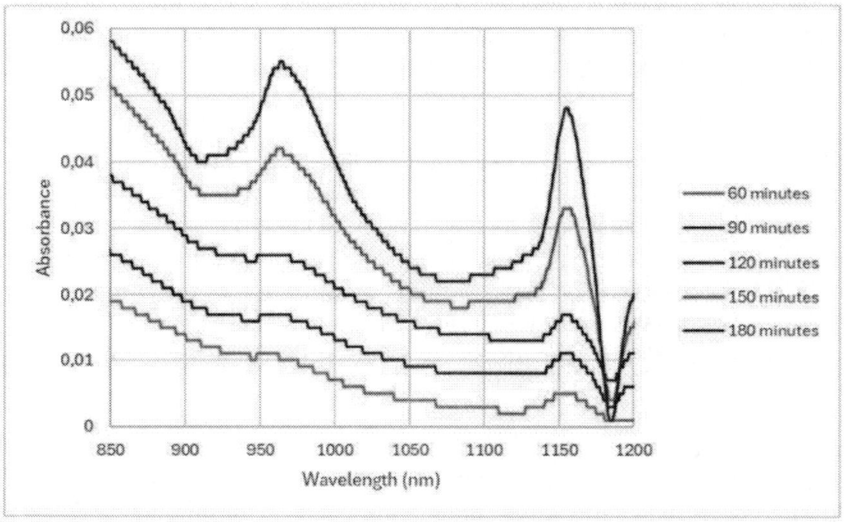

Figure 2. *Phenol ArOH (970 nm) and Methyl/Methylene groups (1150-1160 nm) absorption bands*

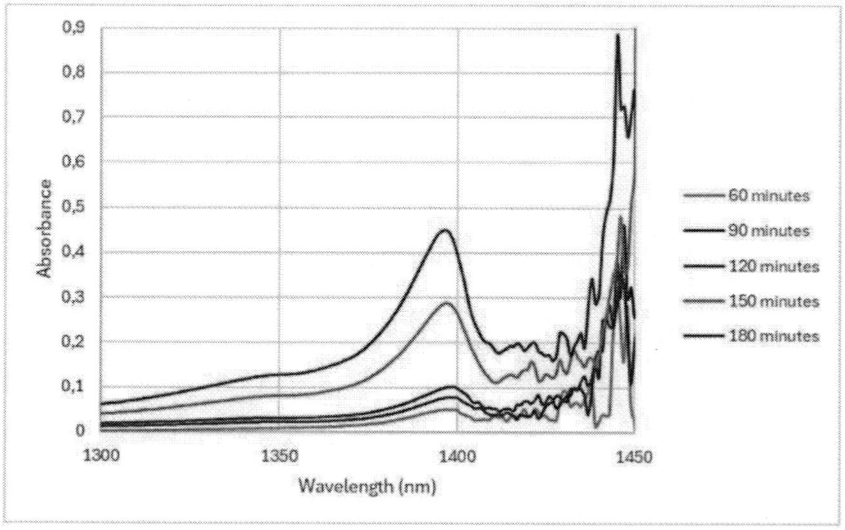

Figure 3. *Absorption band of ROH aliphatic alcohol-type groups and tertiary amines (1395 nm)*

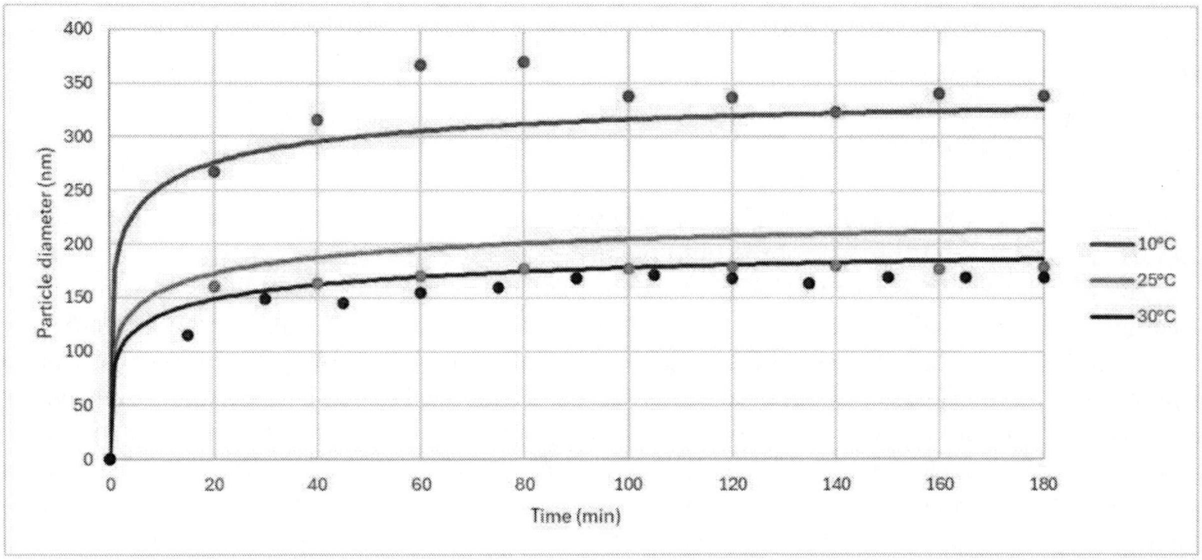

Figure 4. *Behavior of particle diameter as a function of time at different temperatures*

142g Programmable Dynamic Self-Assembled DNA Nanostructures As Linear Actuators

Yanqi Jiang, Rebecca Schulman

Filaments can play key roles in inducing motion within biological cells and systems. For example, actin filaments can exert pushing force onto the plasma membrane through polymerization by their fast-growing end to induce membrane protrusion. Inspired by the actin polymerization, we developed a synthetic, programmable linear actuator using the dynamic growth and shrinkage of DNA nanotubes. This actuator consists of DNA nanotubes with attached capping structures that can be displaced by polymerization or depolymerization. New tiles are labeled by a different fluorophore and epifluorescence microscopy is performed to visualize the insertion displacement. We observed that when the free monomer concentration is higher than the critical value, new tiles are inserted at the nanotube-cap interfaces with flexible connections; no insertion occurs at the nanotube-seed interfaces with rigid interconnects. Reversely, depolymerization happens at the nanotube-cap interfaces with nanotubes getting shortened when the free monomer concentration is lower. Since the kinetics of nanotube tile self-assembly, which drives the displacement, are quite sensitive to monomer concentration, at a specific temperature we can tune the rates of nanotube polymerization and depolymerization by tuning the free monomer concentration. The DNA nanostructures we are building are programmable active elements whose motion could be sensitive to stimuli from the environment by coupling these stimuli to the modulation of monomer concentration. Insertion polymerization could also occur at multiple sites within DNA nanotube chains, so that larger structures could enable faster displacement.

143a Nano-Enabled Intelligent Living Soft Materials for Sustainable Healthcare and Environment Applications

Roya Koshani

An important pillar of sustainable development is the creation of intelligent multifunctional materials derived from renewable resources for real-life applications. Developing innovative nanotechnologies using natural materials and green chemistry may bring us closer to this objective. Within this approach, living soft materials, which encompass cellular and/or acellular components, play a pivotal role. This talk aims to showcase my research endeavors —past, ongoing and potential future— including the nano-/micro-engineering, chemical functionalization, and characterization of biomaterials. The research visions will be structured around two main themes: acellular and cellular living soft materials. First, my expertise in advanced biomaterial development across various platforms encompassing nano-/micro-particles, bulk/granular hydrogels, conventional/Pickering emulsions, cryo-/aerogels, and organo-/emulgels will be presented. Specifically, I will discuss how newly emerged hairy nanocelluloses as one of the most promising sustainable materials of modern times with tunable surface chemistry and unique architecture address persistent healthcare and environmental challenges. Accordingly, new generations of nano-enabled anti-antibiotics to combat the antibiotic resistance in the healthcare sector and highly selective nano-adsorbents for the recovery of globally limited rare earth and precious metals in the environment sector will be introduced. Next, this talk delves into sustainable chemistry and nano-engineering strategies aimed at transforming static soft biomaterial platforms into stimuli-responsive, acellular analogs. In this regard, insights into hybridizing granular and bulk hydrogels via dynamic covalent chemistry to yield extracellular matrix-mimetic and tissue-mimetic platforms will be shared. The last segment of this presentation will cover the specific challenges associated with cell-laden soft materials and explore the novel exciting aspects of integrating living organisms, particularly bacteria and mammalian cells, in a single platform.

143b Formation of Complex Metal Nanostructures By Imprinting Full-Site Addressability from DNA Origami Onto Inorganic Particles

Yifeng Shi, Grigory Tikhomirov

DNA origami – a DNA-based polymer – has been generally considered as a powerful biomaterial to achieve the self-assembly of inorganic plasmonic nanocrystals due to its high programmability and addressability. However, the assembly strategies developed so far are mostly limited to two sub-categories: i) non-finite lattice whose final dimension and shape cannot be controlled and ii) small-scale finite structures with typically only 2-10 nanocrystals. Moreover, in both types of plasmonic nanocrystal-DNA origami hybridization systems mentioned above, the plasmonic nanocrystals are only partially connected to the origami, which limits the number of addressable functionalization sites associated with each nanocrystal. Here we report a new strategy for programmable assembly of gold nanocrystals (spherical particles, rods, and cubes) that can be assembled into multiple complex configurations that are hard to achieve using traditional methods. The assembly of the gold nanocrystals is achieved using a simple rectangular DNA origami. The nanocrystals were first "wrapped" by the origami based on the particle-origami interaction. These wrapped crystals were then assembled into finite lattice with desired configuration through pre-designed single-stranded DNA extensions on the origami rectangles. By adjusting the sequence and position of the extensions, we could correspondingly adjust their assembly patterns. Since almost the whole surface area of the nanocrystals is covered by the origami, these structures distinguish themselves from the previous origami-particle hybridizations through their full-site addressability. Moreover, since the electric field enhancement is proportional to the particle size, the larger nanoparticles applied in the current study could greatly enhance their hotspot effect upon assembly. The well-defined configuration and enhanced hot-spot effect makes this assembly promising in a variety of fields including Surface Enhanced Ramen Spectroscopy (SERS), plasmonic photocatalysis, and plasmonic-enhanced luminescence.

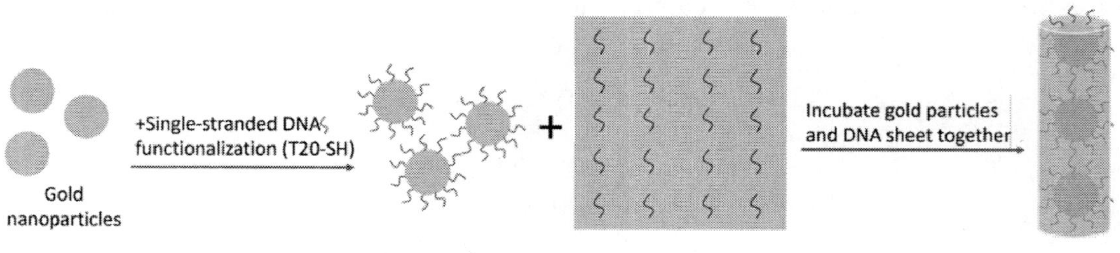

143d Multiplexed Metabolite Detection Using High-Q Metasurfaces and Modular DNA Aptamer Probes

Yuanwei Li, Varun Dolia, Sajjad Abdollahramezani, Parivash Moradifar, Darrell Omo-Lamai, Jennifer A. Dionne

Metabolite detection is crucial for understanding biochemical processes and pathways, and can facilitate more accurate and timely therapeutic intervention. For example, the metabolites cortisol, creatinine, ATP, and adenosine are associated with chronic stress, kidney dysfunction, and cardiovascular disease, but as of yet, no point-of-care test exists to quantitatively monitor their presence and abundance. Nanophotonic biosensors provide a platform for tracing these signature biomarkers, enabling precision health, early disease diagnosis, staging, and monitoring. Existing sensors based on plasmonic nanostructures face challenges such as limited sensitivity for low-concentration analytes and a lack of multimodal sensing capabilities.

Here, we demonstrate sensitive, multiplexed detection of these metabolites by integrating high-quality (high-Q) dielectric metasurfaces with spherical nucleic acid (SNA) reagents. Specifically, we design DNA sequences (aptamers) that undergo a structural change upon target metabolite binding. A split portion of this aptamer is bound as a single-strand to the SNA nanoparticle core (typically ~10nm diameter gold). Another portion of this split aptamer is bound to our resonant silicon nanophotonic antennas, termed VINPix[1], which can have quality factors in the thousands to tens of thousands. Upon the presence of target analytes, the DNA sequences on the metasurfaces and the SNAs interact with the target analyte molecules, resulting in the formation of a secondary DNA structure. This interaction effectively localizes the analytes and SNAs on the dielectric metasurfaces and creates localized hotspots, leading to a significant, visible-detectable resonance shift. Concurrently, the SNAs' attachment to the high-Q metasurfaces results in a visible color change of the metasurfaces, providing a visual signature of the biomarkers' presence. We demonstrate the detection of clinically-relevant, nanoMolar concentrations of cortisol, creatinine, ATP, and adenosine. Additionally, we show how our modular VINPix arrays and SNA structures can be adapted to different biomarkers, for targeted and multiplexed detection of other molecules, including proteins, microRNAs, and ions. Finally, we show multiplexed detection of metabolites and proteins in clinical samples of urine. By quantifying multiple biomarkers in urine, our integrated approach provides vital health metrics like the urine albumin to creatinine ratio (uACR), a key indicator for kidney disease monitoring.

Reference:

1. Dolia, V., Balch, H., Dagli, S., Abdollahramezani, S., Delgado, H. C., Moradifar, P., ... & Dionne, J. A. (2023). Very-Large-Scale Integrated High-Q Nanoantenna Pixels (VINPix). arXiv preprint arXiv:2310.08065.

143e Dye-Mediated Reversible Fluorescence Quenching for SWCNT-Based Nanosensors

Shoichi Nishitani, Ashvin Irrinki, Ava Moudi, Ankita Biyani, Feiyang Chen, Padma Muppirala, Kevin Ao, Markita Landry

Single-walled carbon nanotubes (SWCNTs) are attractive materials for biochemical sensing and imaging, due to their stable photoluminescence in the tissue-transparent, near-infrared (NIR) region. Certain single-stranded DNA (ssDNA) biopolymers have been known to effectively quench the fluorescence of SWCNTs, a process made reversible by the introduction of biomolecules such as catecholamines. These nanosensors have enabled visualization of dopaminergic neuromodulation, yet with limitations in 1- selectivity of dopamine over other catecholamines, and 2- the necessity to screen for molecular recognition. These limitations motivated us to devise nanosensors with the capability to detect a broader spectrum of biomolecules based on an engineered quenching phenomenon.

Apart from ssDNA, certain redox dyes can quench SWCNT fluorescence via surface binding. Excitingly, we find that catecholamines do not reverse redox dye-based quenching, instead, fluorescence de-quenching can be selectively reversed upon removal of the dye molecules from the SWCNT surface. Motivated by this finding, we sought to design and develop rationally-designed, reversible dye-quenching mediated nanosensors capable of selectively detecting target biomolecules.

As a proof-of-concept, we designed the SWCNT corona where DNA hybridization is linked to the dynamic movement of methylene blue (MB), a representative dye molecule used in this study, on the SWCNT surface; thus, selectively modulating SWCNT fluorescence emission. In particular, $(GT)_{15}T_{20}$-wrapped SWCNTs were designed as a model to hybridize with the target sequence, A_{20}. To show that the movement of MB on the surface modulates SWCNT fluorescence, we designed multiple variations, where MB was conjugated either to the 3'-end or between $(GT)_{15}$ and T_{20} of $(GT)_{15}T_{20}$, or to the 5'-end or 3'-end of A_{20}. All ssDNA-wrapped SWCNTs were prepared by general sonication-based dispersion. Importantly, we prepared these nanosensors under conditions where ssDNA does not quench the fluorescence; thus, eliminating any contribution from non-specific turn-on responses.

When MB was conjugated to $(GT)_{15}T_{20}$, fluorescence intensity specifically increased upon incubation with A_{20}. As expected, the increment was significantly stronger when MB was conjugated to the 3'-end of the sequence than between $(GT)_{15}$ and T_{20}, because the hybridization does not affect the position of MB in the latter case. For the other design where MB was conjugated to A_{20}, A_{20}-MB (MB conjugated to 3'-end) more effectively quenched the fluorescence of $(GT)_{15}T_{20}$-SWCNT than MB-A_{20}, which was also expected from the unidirectional nature of DNA hybridization. Overall, these results strongly suggest that the movement of MB away from the SWCNT surface modulates SWCNT fluorescence intensity. Notably, the normalized change in fluorescence intensity reached 100 % upon optimization for $(GT)_{15}T_{20}$-MB-SWCNT. These nanosensors' high turn-on ratio can be beneficial in developing NIR fluorescence nanosensors for DNA and RNA detection, and possibly in the future for other biomolecules. Correspondingly, we used our design to develop nanosensors capable of detecting viral RNA fragments in plants, using tobacco mosaic virus as a model pathogen. Thus, our results demonstrate the potential of using redox-dye movement relative to the SWCNT surface as a fluorescence mediator to develop rationally-designed, selective nanosensors for the detection and imaging of a broad range of biomolecules.

143f Wavelength Tunable Infrared Perfect Absorption in Plasmonic Nanocrystal Monolayers

Woo Je Chang, Zarko Sakotic, Alexander Ware, Allison Green, Benjamin Roman, Kihoon Kim, Thomas Truskett, Daniel Wasserman, Delia Milliron

The ability to efficiently absorb light in ultrathin (subwavelength) layers is essential for modern electro-optic devices, including detectors, sensors, and nonlinear modulators. Tailoring these ultrathin films' spectral, spatial, and polarimetric properties is highly desirable for many, if not all, of the above applications. Doing so, however, often requires costly lithographic techniques or exotic materials, limiting scalability. Here we propose, demonstrate, and analyze a mid-infrared absorber architecture leveraging monolayer films of nanoplasmonic colloidal tin-doped indium oxide nanocrystals (ITO NCs). We fabricate a series of ITO NC monolayer films using the liquid–air interface method; by synthetically varying the Sn dopant concentration in the NCs, we achieve spectrally selective perfect absorption tunable between wavelengths of two and five micrometers. We achieve monolayer thickness-controlled coupling strength tuning by varying NC size, allowing access to different coupling regimes. Furthermore, we synthesize a bilayer film that enables broadband absorption covering the entire midwave IR region ($\lambda = 3–5$ μm). We demonstrate a scalable platform, with perfect absorption in monolayer films only hundredths of a wavelength in thickness, enabling strong light–matter interaction, with potential applications for molecular detection and ultrafast nonlinear optical applications.

144a Efficient and Cost-Effective Catalytic Materials for Upgrading of Carbon Dioxide

Jingguang G. Chen

Rising atmospheric concentration of CO_2 is forecasted to have potentially disastrous effects on the environment from its role in climate change and ocean acidification. Converting CO_2 into value-added chemicals and fuels is one of the most practical routes for reducing CO_2 emissions while fossil fuels continue to dominate the energy sector. One of the challenges is the development of efficient and cost-effective catalysts for the thermochemical and electrochemical conversion of CO_2. In this talk we will discuss the design and evaluation of catalytic materials for CO_2 upgrading using a combination of kinetic studies, *in situ* characterization and density functional theory calculations. We will present several examples of (1) CO_2 conversion by thermocatalysis, (2) CO_2 reduction by electrocatalysis, and (3) CO_2 upgrading using tandem electrocatalysis-thermocatalysis. We will use these examples to highlight the importance of using fundamental chemical engineering principles to guide the selection of effective catalysts.

144b Designing Better Materials and Systems for Future Batteries

Y. Shirley Meng

High energy long life rechargeable battery is considered as the key enabling technology for deep de-carbonization. Energy storage in the electrochemical form is attractive because of its high efficiency and fast response time. Besides the technological importance, electrochemical devices also provide a unique platform for fundamental and applied materials science & research since *ion* movement is often accompanied by inherent complex phenomena related to phase changes, electronic structure changes and defect generation. In this plenary talk, I will discuss a few new perspectives for energy storage materials including new superionic conductors, reactive metal anodes and their interfacial engineering in a device and system. With recent advances in advanced photon and electron characterization tools and computational methods, we are able to explore ionic mobility, charge transfer and phase transformations in electrode and electrolyte materials *in operando*, and map out the structure-properties relations in novel functional metals, ceramics and gaseous materials for next generation energy storage and conversion. Moreover, I will discuss a few future priority research directions for electrochemical energy storage.

144c Cool Plastics for Energy Sustainability

Natalie Stingelin

With seabirds trapped in multipack drink rings, and mid-ocean islands of indestructible rubbish, the idea that plastics could play a big part in a sustainable future world might seem far-fetched. However, new smart plastics may yet rescue the reputation of this all-consuming 20th century material. Research into 'cool plastics' for cars and buildings could reduce the need for air conditioning and, thus, improve their energy efficiency. We will present recent efforts to design plastics of desired optical functions targeted for a greener world. We will discuss the potential of such systems that can offer the same flexibility, softness and light weight as commodity plastics but can control the flow of light and heat therefore assisting energy management in buildings and greenhouses in the form of heat mirrors, photovoltaic applications when used as anti-reflection coatings and semi-transparent mirrors, as well as building blocks for novel optical structures that can lead to quantum devices.

182b Silver Nanoparticle Sensor Array for the Detection of Sars-Cov-2 for Nanoscale Science and Engineering Forum

Benjamin Lam

We report a silver nanoparticle (AgNP) sensor array to detect SARS-CoV-2 viral particles utilizing five peptide receptors to produce a colorimetric response. We show that the interaction of the virus particles with the peptides interferes with the shift in plasmonic absorbance and hinders the formation of snowflake-like fractal assemblies between the AgNPs and the bridging peptides. The limit of detection in water of the sensor array was 500,000 viral copies/mL. We demonstrate the capabilities of the sensor array to distinguish between SARS-CoV-2 (Beta, Delta, and Omicron variants) from the influenza virus at the 99% confidence level through linear discriminant analysis. The sensor array was also able to discriminate three out of five negative exhaled breath condensate samples from five positive SARS-CoV-2 EBC samples.

182c Fluorescent Nanothermometry-Enabled Robotic Minimally Invasive Laser Tissue Soldering

Oscar Cipolato, Inge Herrmann

Laser tissue soldering combines laser light with a specialized adhesive material called solder paste, offering a compelling alternative to traditional sutures and staples. This innovative surgical technique primarily utilizes protein-based adhesives to form robust, watertight bonds. Its suitability for minimally invasive procedures presents distinct advantages over conventional methods, particularly in delicate tissues like blood vessels, nerves, and the intestine, where complications such as inflammation, infection, and fluid leaks can arise.

However, the effectiveness of laser tissue soldering heavily relies on precise temperature management during the procedure. Our research introduces a novel approach to address this challenge, employing controlled laser tissue soldering guided by fluorescent nanothermometers. These nanothermometers, seamlessly integrated into protein-based solders, enable non-invasive temperature monitoring. As a significant step toward more accurate and less invasive soldering techniques, we demonstrate automatic temperature control and intelligent laser power modulation for achieving optimal outcomes. This progress is facilitated by incorporating an intelligent solder paste containing near-infrared (NIR) fluorescent nanothermometers and nanoabsorbers such as gold nanorods or titanium nitrate nanoparticles.

Moreover, our study integrates computer vision-based solder recognition and various strategies to guide surgery by analyzing the temperature distribution based on nanothermometry during soldering, thereby enhancing procedural precision. We also emphasize the crucial advantages of minimally invasive soldering in clinically relevant scenarios.

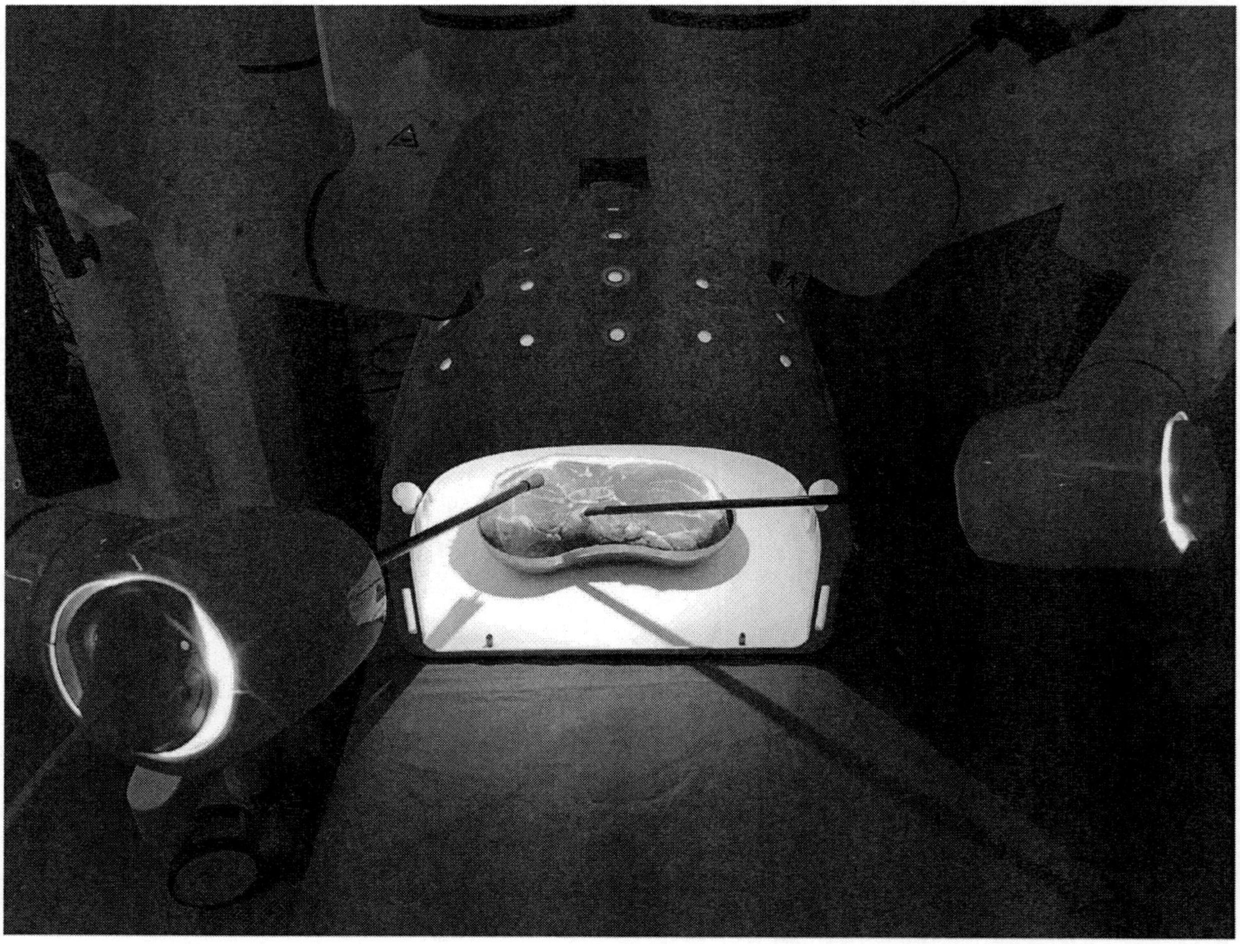

182d Glycan-Modified Cellular Nanosponges for Enhanced Neutralization of Botulinum Toxin

Dan Wang

Botulinum toxin (BoNT) is a potent protein neurotoxin that disrupting neural transmission by inhibiting the release of neurotransmitter. The toxin has been found in medical applications. Its characteristics make it a potential bioterrorist weapon, which threatens public health. Currently, there is a lack of effective antidote for BoNT in the event of a biological attack. Here, we propose a novel solution with cell membrane coating nanotechnology. The glycan modified cellular nanosponges, developed via metabolic engineering, acted as cell decoys to bind with BoNT for detoxification. To prepare the nanosponges, we first increased the expression levels of gangliosides on THP-1 cells through metabolic engineering. Then, nanosponges were made by coating the modified THP-1 cell membrane onto synthetic polymeric cores. Our findings showed that increased levels of gangliosides were expressed on the cellular nanosponges, leading to higher binding capacities with BoNT. Therefore, the glycan-modified cellular nanosponges exhibit superior efficacy in neutralizing BoNT in a synaptic vesicle recycling assay when compared to their unmodified counterparts. In a mouse model of BoNT intoxication, the glycan-modified cellular nanosponges showed survival benefits when administered both as a treatment and a preventative regimen. These results highlight the potential of cellular nanosponges, especially when modified with glycans, as a promising platform against BoNT and related clostridial toxins.

182e Enhancing Light Absorption Properties of Hybrid Magnetic Nanoparticles Using Metal Chelation and Electroless Deposition of Metal Ions.

David Mier Salazar, Roberto Guzman

This work describes the different characteristics of light absorption of hybrid magnetic nanoparticles when using different chelated metal ions or electroless deposited. Fe_3O_4@Agarose core-shell nanoparticles were modified to incorporate polydopamine (PDA) as a second shell. Fe_3O_4@Agarose-PDA nanoparticles make black solutions capable of absorbing a considerable amount of light. However, to make the nanoparticles more conductive, a metal can be deposited onto the surface of the particles. In this work, metal chelation and electroless deposition were proposed as a way to increase conductivity and change the absorption characteristics of the particles.

182f Toward a Rational Design of Engineered Nano-Photonic Threshold Scale Inhibitors

Nadhem Ismail, Ali Alshami

Scaling, the formation of sparingly soluble salts such as calcium carbonate, is a persistent problem in several water-based industries. The use of chemical scale inhibitors has been at the forefront of scale control strategies. The use of emergent nano-photonic (Quantum Dots) scale inhibitors has unfolded in the last few years owing to their stable fluorescence, promoted scale inhibition efficacy by the high surface area to volume ratio (A/V), and facile tailoring via molecularly engineering the surface. Negatively charged functional groups such as carboxylates, sulfates, and phosphates interfere with the nucleation and crystals growth of inorganic minerals and destabilize the formed polymorphs in the early stages in what is called the threshold effect. Despite the substantial efforts in developing new nano-photonic scale inhibitor platforms, most of the developed platforms report on carboxyl functionalized platforms with no mechanistic study on how the synergy between the functional groups and molecular conformation affects the nano-photonic scale inhibitors performance. In this study, we aim to broaden the scope of our previously reported results on Carboxyl Silicon Quantum Dots (CSiQDs) to include different quantum dots core (i.e. Carbon Quantum Dots) and different functional groups on the core platforms. We aim to study the relationship between the nano-photonic scale inhibitor antiscaling properties and the functional groups, solubility, flexibility, and fluorescence properties. The fluorescence properties will facilitate the inhibition mechanism elucidation of the different functional groups through confocal imaging. This work may pave the way for the rational design of subsequent generation of nano-photonic scale inhibitors.

182g Silver Nanoparticles Immobilized on Heat Treated Candle Soot for the Efficient Reduction of Aqueous Nitroarenes and Azo Dyes

Sourav Halder, Bhaskar Bhaduri

The imprudent use and disposal of toxic chemicals into the environment in the name of global industrialization is creating a massive ecological crisis. Toxic nitroaromatic compounds like *para*-nitrophenol (4-NP), *para*-nitroaniline (4-NA) and *meta*-dinitrophenol (2,4-DNP) are used as precursors across various industries for the production of pharmaceuticals, pesticides and fungicidal agents, dyes and rubber, explosives and industrial solvents. Synthetic azo dyes like Methylene Blue (MB) and Congo Red (CR) are found in effluents of industries like textiles, paper and pulp and paints and varnish. Dyes and nitro-substituted organic compounds are carcinogenic and cytotoxic to humans and can cause dermal irritation, kidney inflammation and liver cirrhosis. Due to the presence of a nitro group, nitroaromatics are highly water soluble, recalcitrant and stable towards biodegradation and other conventional wastewater treatment methods. Noble metal-catalysed reduction of nitroaromatic compounds to their corresponding environmentally benign aniline derivatives is tone of the only useful and viable routes for environmental remediation of nitroarenes.

A facile in-situ method for preparing stable Ag nanoparticles (average diameter ≈28 nm), decorated on heat-treated candle soot (cs-HT) is hereby reported. The synthesised Ag@cs-HT nano-catalyst with ~2% wt./wt. Ag loading was thereafter investigated for its effectiveness towards catalytic reduction of aqueous nitroarenes like 4-NP, 4-NA and 2,4-DNP and azo dyes like MB and CR using sodium borohydride ($NaBH_4$) as the reducing agent under ambient conditions. The effect of various process parameters like $NaBH_4$ dosage, pH and temperature on the efficiency of reduction were investigated. The catalytic reduction followed pseudo first-order kinetics for all nitroarenes and dyes, with 4-nitrophenol achieving the highest degree of reduction (~99.96%) within 5 minutes at 35 °C, solution pH 3 and an optimized nano-catalyst dose of 0.08 g/L and $NaBH_4$ dosage of 1.8 mM. The detection of $H^{\bullet}$ radical adducts indicates that Ag nanoparticles adsorb BH_4^- ions from solution to form Ag-H species. The enhanced removal capability of Ag@cs-HT was attributable to the high specific surface area and hydrophilicity of cs-HT (S_{BET}=152.3 m^2/g, average contact angle=23.4°) which facilitates rapid adsorption of BH_4^- ions and reactant molecules, followed by electron transfer from the active Ag-H complex and subsequent reduction of the adsorbed reactant species. The presence of anions like SO_4^{2-}, CO_3^{2-} and NO_3^- offered marginal resistance to the reduction process, except for Cl^- which anomalously enhanced it, and PO_4^{3-} which drastically reduced the reduction efficiency. The synthesized Ag@cs-HT nano-catalyst was found to reusable for at least 8 consecutive cycles for reduction of 4-nitrophenol, without significant loss of catalytic activity and minimal leaching of Ag. The aforementioned outcomes clearly establish Ag@cs-HT as an effective catalyst for removal of highly toxic nitroarenes and azo dyes from aqueous solutions.

182h Uniformly Dispersed FeTe2 Nanoparticles on Electrospun Nanofibers As Free-Standing Anodes for Sodium Ion Batteries

Xing Li, Zhi Liu, Zhen Wei, Ling Fei

Sodium-ion batteries (SIBs) are increasingly being seen as a promising alternative to traditional lithium-ion batteries (LIBs), largely due to the abundance of sodium reserves and the high safety and relatively low cost of the material used in their production. However, like their lithium-ion counterparts, the anode materials in SIBs face a number of challenges that must be overcome before they can truly live up to their potential. Two key issues currently hindering the development of effective anode materials are sluggish electrochemical kinetics and a tendency for large volume expansion during the charge/discharge cycle. These factors not only impact the overall performance of the batteries, but also present significant challenges in terms of their long-term durability and reliability.

Tellurium, a metalloid with unique physical and chemical properties, exhibits exceptional stability and conductivity, making it an ideal material for a wide range of applications. Similarly, carbon fiber, renowned for its remarkable flexibility and conductivity, has emerged as a promising alternative to conventional current collectors. In this study, we successfully fabricated FeTe2@CNFs, a novel composite material, by employing a simple yet cost-effective approach. The method involves electrospinning a mixture of Te powder and specific precursors, followed by annealing at a controlled temperature. The result is a uniform dispersion of FeTe2 nanoparticles within the carbon fibers, a feat achieved through in-situ growth. These nanoparticles play a pivotal role in enhancing the overall capacity performance. The choice of iron and tellurides for this study was influenced by several factors. Iron, with its low cost and availability, was a logical option. On the other hand, tellurides, like their sulfide and selenide counterparts, offer superior conductivity and stability, making them ideal candidates for the development of novel materials. The distinctive CNFs structure, characterized by its high conductivity and vast exposed surface.

182i Density Functional Study for Atomically Dispersed Noble Metal on Mesoporous Carbide By Topotactic Conversion

Hayoung Jeong, Jeong Woo Han

The highly ordered mesoporous Mo_xC synthesized through the Kirkendall effect and the catalytic effect of nanomaterials is a key factor in the nanostructuring of Mo_xC. Catalytic effects have been observed with noble metals such as Pt, Ru, and Au. Notably, in the presence of noble metals, topotactic conversion is facilitated. Molybdenum carbide exists in two forms: α-MoC and β-MoC. The formation of α-MoC occurs via topotactic conversion induced by noble metal catalysis. Moreover, mesoporous Mo_xC predominantly exhibits the α-MoC phase. To elucidate the topotactic conversion mechanism and noble metal catalytic effect on two types of Mo_xC, density functional theory (DFT) calculations were conducted. Through investigation into defect formation mechanisms occurring via different pathways—α-MoC made from initial MoO3 through MoO_xC_y and β-MoC made from MoO_2—the process of topotactic conversion is elucidated. Identifying the preferred vacancy site and investigating the possibility of migration when generated in less favored locations, it focuses on analyzing the pore structure formed during MoC synthesis and the process of carbon substitution. Analyzing the entire process allows for the identification of defect formation locations and elucidation of the overall pore formation mechanism, enabling an explanation of the Kirkendall effect. Additionally, comparisons of electrochemical performance were made through calculations of the hydrogen evolution reaction, water dissociation, and hydrogen spillover. In the case of α-MoC, it was observed that there were better performances in HER and water dissociation. Additionally, it was confirmed that the performance was further enhanced with the addition of Pt. Finally, a comparison between Phosphomolybdic acid, utilized as the precursor for Mo, and H_2PtCl_6 or platinum acetylacetonate ($Pt(acac)_2$), employed as the precursor for platinum, is conducted to assess the interactions of these compounds. Through this comparison, the objective is to evaluate the two precursors. Specifically, $Pt(acac)_2$'s interaction with PMA results in a higher Pt dispersion, thereby preventing the formation of Pt agglomerates. Subsequently, interaction energy calculations are performed to highlight the differences between the precursors and compare the outcomes with calculations involving other noble metals instead of Pt. The findings of this study demonstrate the advantages of α-MoC's topotactic conversion and are expected to provide insights into the distinguishing features of α-MoC compared to other MoC forms.

182j Growth Mechanism Study of Metal-Organic-Framework Encapsulated Laccase Using E-QCM-D

Jingyi Li

Enzymes, revered for their efficiency, specificity, and mild reaction conditions, play pivotal roles in biosensing, detection, and drug delivery. Embedding enzymes into metal-organic-framework (Enzyme@MOF) not only enhances the intrinsic fragility of enzymes but also imparts bioactivity to the MOF structure. Despite exhibiting good stability, Enzyme@MOF complexes are limited in their biological functionality. Due to the rigid MOF structure and the interactions between enzymes and metal nodes of MOF, potentially affecting the bioactive conformation. Therefore, fundamentally understanding the growth mechanism of Enzyme@MOF and the behavior of biomolecules during the preparation process is crucial for highly active Enzyme@MOF complexes. In this work, we used the electrochemical quartz crystal microbalance with dissipation (E-QCM-D) to correlate MOF structure growth and changes in bioactivity, exploring the growth mechanism of Enzyme@MOF and their bioactivity transformations during the early encapsulation process. We established a current-based laccase enzyme activity assessment method using ABTS as the substrate, to investigate the growth mechanism of zeolitic imidazolate framework (ZIF) encapsulated laccase complexes. We studied the interaction between laccase and precursors and its effect on enzyme activity by variational concentration of metal ions (1, 10, 40 mM), types of organic ligands (2-methylimidazole, 1-methylimidazole), and their concentrations (4, 40, 160 mM). Frequency and current signals revealed that electrostatic interactions significantly influenced bioactivity, with metal ions enhancing bioactivity while causing minimal mass loading fluctuations. Notably, a significant mass increase within the first 120 s indicated in-situ co-precipitation of Laccase@ZIF complexes during the mixed precursors. Furthermore, by observing and analyzing the current changes in the early 120 s of Laccase@ZIF preparation, we explored the key factors influencing the transformation of bioactivity during the Laccase@ZIF preparation process. To illustrates the growth mechanism, this study focused on two aspects: the interaction between laccase and precursors and the changes of bioactivity, which provides critical insights for further developing tailored Enzyme@MOF complexes with enhanced functionalities.

183a Near-Infrared Circularly Polarized Luminescence-Active Semiconductor Helices with Tunable Twist Counts

Kody Whisnant, Jun Lu, Nicholas A. Kotov

Circularly polarized luminescence (CPL)-active materials have recently gained increasing interest due to their potential applications in 3D displays, optoelectronic devices, optical sensors, and optical information storage. Circularly polarized light, generated by passing light through a linear polarizer and quarter-wave plate, is an essential component of various optical techniques and devices. CPL in the NIR has been found to have advantages over visible light for the areas of bioimaging and telecommunications, resulting from the low absorption of NIR light in biological samples allowing for high-quality imaging with minimal sample damage and low attenuation of NIR light in optical fibers which is ideal for long-distance communication. Few cases of NIR-CPL active materials along with limited characterization equipment for CPL detection in the NIR has however made the expansion of this material library challenging. This work seeks to address these challenges using mercury-exchanged cadmium telluride (CdTe) mesoscale helices that display NIR-CPL activity.

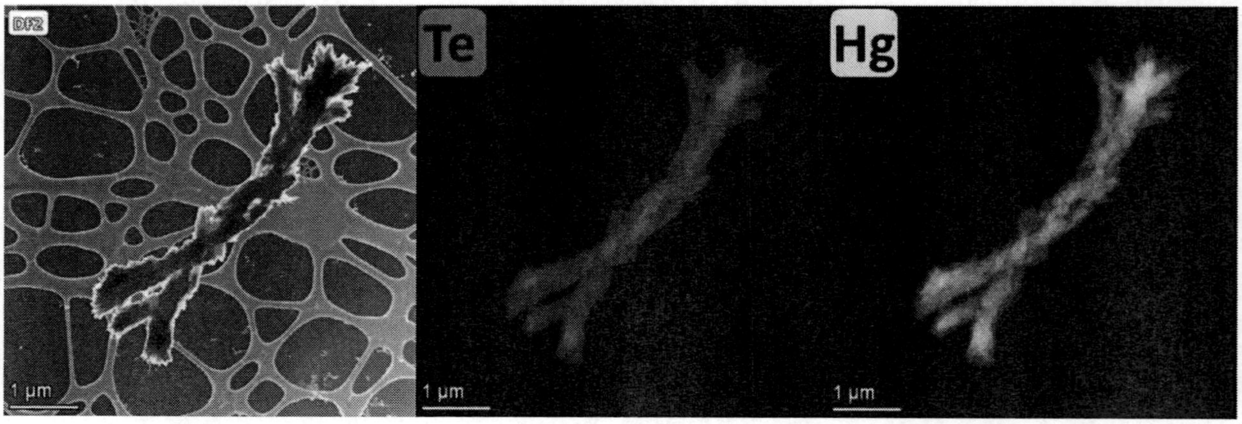

183b Poster: Designing the Fluorinated Superfine-Activated Carbon Adsorptive Membrane for Selective Removal of Long- and Short-Chain PFAS

Medha Kasula, Milad Esfahani

Membrane filtration and adsorption are two prominent methods for the effective removal of per- and poly-fluoroalkyl substances (PFAS) from water. Thin-film composite (TFC) nanofiltration (NF) membranes exhibit acceptable PFAS removal (<90%), but the presence of rejected PFAS in the concentrate stream necessitates an additional separation process. Our previous study reported the fabrication of adsorptive TFN NF membranes, incorporating superfine activated carbon (SFAC), for enhanced perfluorooctanesulfonic acid (PFOS) removal via adsorption during water filtration. The SFAC-TFN-0.5 membranes (0.5 wt% loading) of SFAC embedded inside the polyamide layer of TFN demonstrated 94% PFOS removal, with over 80% adsorption, and an increased permeability of 22 LMH/bar compared to the bare TFC. Although the results were impressive, the selectivity of the membranes towards different PFAS (long-chain PFOS and short-chain perfluorobutane sulfonate, PFBS), and in the presence of other components was not investigated. In this project, we investigated the selectivity of the adsorptive membrane toward different PFAS in the presence of NOM. Initial static adsorption results revealed limited PFBS adsorption on SFAC (45%), while it showed 93 % adsorption towards PFOS. To enhance the adsorbent's selectivity for different PFAS, SFAC was further fluorinated, as fluorination increases fluorophilic interactions with short-chain PFAS — an essential mechanism for their adsorption. The selectivity of PFOS in the presence of NOM on the SFAC membrane was observed to decrease to 70 % from 90 %. Therefore, the hypothesis is that fluorophilic interaction can boost the affinity of SFAC towards PFOS rather than NOM. In addition, the incorporation of SFAC in the IP process led to SFAC aggregation and disruption of the polyamide layer as reported in our previous study. In this project, fluorinated SFAC is grafted onto the active layer to prevent aggregation and better control surface properties.

183d Evaluating Performance Variations in Ti3C2Tx Mxene Composite Polymer Electrolytes for Solid State Batteries: Analysis of Different Processing Techniques

Sahand Serajian, Mona Bavarian

This work studies the incorporation of Ti3C2Tx MXenes into polymer electrolytes for addressing the issue of low ionic conductivity in solid electrolytes. Herein, the effect of two distinct MXene integration methods, vacuum filtration and freeze-drying, will be compared. The electrochemical impedance spectroscopy and differential scanning calorimetry are employed to study the impact of MXene integration on the ionic conductivity and mobility within the polymer matrix. Remarkably, the findings revealed that freeze-dried MXene not only significantly reduces the crystallinity of the SPE by **4.9%** but also elevates the ionic conductivity by **an order of magnitude** compared to its vacuum-filtered counterpart. The influence of material processing techniques on ionic conductivity was studied by equivalent circuit modeling and voigt circuit representation models. This study indicates the synergistic effects of MXene composites in energy storage applications, thereby unraveling the impact of processing conditions of this class of 2D materials in energy storage applications.

183e Nanoscale Mechanical and Morphological Characterization of Ebolavirus-like Particles: Implications for Therapeutic Development

Hannah Hargrove, Sajedeh Yazdanparast, Xiaoji Xu, X. Frank Zhang

Abstract

Ebolavirus (EBOV) is one type of filovirus that causes the deadly EBOV disease, with an average fatality rate of around 50%. EBOV outbreaks are devastating and unpredictable and may emerge as the next global pandemic. As a BSL-4 pathogen, EBOV is inaccessible to regular biological laboratories. Therefore, EBOV virus-like particles (EBOV-VLPs) and EBOV pseudoviruses (EBOV-PVs) are utilized in initial development of many potential therapies, as opposed to using live viruses. These particles must accurately represent the morphological, mechanical, and biochemical properties of live EBOVs. Due to EBOVs' nanometer size and irregular, filamentous shape, these properties have been challenging to characterize. In this research, state-of-the-art nanoscale characterization techniques were employed to examine surface characteristics of a selection of commonly used EBOV-approximating particles. In summary, the study comprehensively determined the accuracy of EBOV approximation with virus-like particles and pseudovirus, and the uniformity of mechanical, structural, and biochemical traits across different EBOV approximations. This provides important implications for developing therapeutic treatments against EBOV using these approximating reagents.

1. Methods

Atomic force microscopy (AFM) was used to characterize each of the EBOV-approximating particles. Each particle was observed hydrated under 1x phosphate buffer serum (PBS), to approximate the particles' *in-vivo* state. During analysis, three protein solutions were used. The UV-inactivated EBOV (UV-EBOV) and the EBOV pseudovirus (PV) stock solutions were supplied by Wendy Maury at Iowa State University, and the EBOV-like particle (VLP) stock solution was supplied by IBT Bioservices. Each of these particles mimic membrane characteristics of EBOV, so they are well-suited to comparative study.

1.1. Peakforce AFM

Each particle was examined via Peakforce AFM imaging mode. This mode is owned by Bruker AFM and utilizes a variation of tapping mode which averages a certain number of force scans per pixel to construct a topographical image of the surface of the sample. By this method, individual nanoscale particles can be imaged topographically—this is how physical conformation of each particle was determined.

Another function of Peakforce AFM scanning is nanoindentation. By this method, force scans can be conducted precisely over the location of a particle based on location coordinates determined by a Peakforce AFM image scan. This force scan registers a small indentation in the surface of the particle, and by analyzing force and depth of the nanoindentation, elastic modulus of the particle can be established. To minimize interference from the substrate, nanoindentation depths of 1 to 5 nm were used in analysis. Any indentations deeper than this are assumed to be affected by the stiffness of the substrate and have been discarded from the dataset.

1.2. Adhesion Force Mapping

Adhesion force mapping is another force scanning AFM functionality, whereby an AFM cantilever is treated with an antibody particle, which is then contacted with a particle of interest adhered to a stiff surface. The force which is required to separate the tip from the surface is then analyzed to determine the binding affinity between the antibody and particle of interest. In this study, adhesion force measurements are taken over a single-particle sample, to distinguish between adhesion on the particle and adhesion over the surface. This allows negative control of each force map to be the substrate background, and sample adhesion can be thus compared to the negative control. As UV-EBOV is the closest approximation to live

EBOV, adhesion force between UV-EBOV and antibody-treated tip will be regarded as the positive control for comparison between adhesion mapping of different particles. For specific sample preparation and testing protocols, see Supplemental Methods.

2. **Results**

2.1. Ebola Virus-Like Particles

Although UV-EBOV and EBOV PV are known to be similar to live EBOV, neither is commercially available. For this reason, the EBOV-VLP is highly useful. The VLP possesses the outer membrane characteristics and some of the inner particle characteristics, but lacks the ability to self-replicate, making it safe for handling in a BSL-2 lab [1]. To determine whether this is a good approximation of live EBOV, the VLP was characterized by the same methods which the UV-EBOV and PV were characterized, and then compared to the UV-EBOV characterization as the most EBOV-like particle.

Peakforce AFM imaging of VLP particles on a treated mica surface revealed that the average width of the VLP particle when hydrated in PBS was 115.1 ± 55.5 nm, the average length of the particle was 480.2 ± 152 nm, and the average height was 51.5 ± 18.9 nm. Overall, the particles exhibited a cylindrical or pill shape. A sample Peakforce scan of the VLP particles adhered to a treated mica surface can be observed in Figure 2.

When hydrated in PBS, the VLP had an average elastic modulus of 8.54 ± 1.69 MPa. The histogram of elastic modulus data is depicted in Figure 3.

Another important consideration when characterizing the VLP particle is determining whether the particle has PS on the surface of the membrane, like live EBOV does. This was assessed using Adhesion Force Mapping, which characterized adhesion forces between a Bruker MLCT probe treated with Annexin V (A5), and the VLP. The resulting force map can be seen in Figure 4.

2.2. Ebola Pseudovirus

While the closest allegory to live EBOV is UV-EBOV, it is not always feasible for a group to obtain. At the time of writing, UV-EBOV is commercially unavailable, and must be obtained from a BSL-4 lab. If a lab cannot do this, another commonly used approximation of EBOV is Ebola pseudovirus (PV), which is formed by expressing EBOV glycoproteins on the surface of a more benign virus. To assess EBOV PV as an approximating particle, our lab obtained a sample from our collaborator of vesicular stomatitis virus (VSV) which has had EBOV GP artificially implanted in the surface of the virus to mimic EBOV's ability to enter the endosome. The PV was then tested and in the coming weeks will be analyzed against the characteristics of UV-EBOV.

When characterizing the PV, AFM Peakforce tapping mode revealed that the PV particles exhibited an oblong pill shape, with a domed top. The average width of the particle was 70.4±20.3 nm, the average length was 285±74.6 nm and the average maximum height of the particle was 7.4±4.2 nm. Figure 5 parts (a), (b) and (c) demonstrate the distribution of height, length, and width measurements of individual particles, respectively. Peakforce nanoindentation of a PV-saturated mica substrate revealed that the average elastic modulus of the PV particle was 0.64 MPa, with a median of 0.49 MPa, a geometric mean of 0.47 MPa, and a standard deviation of 0.51 MPa. Figure 5 part (d) depicts the distribution of elastic modulus measurements across several hundred force curves. It is of note that while the histogram has a left-leaning distribution, there is a distinct peak and then fall to the histogram data, which indicates that the peak of the histogram is interpretable, and the elasticity is large enough to be measured by Peakforce nanoindentation methods.

To assess the presence of phosphatidylserine (PS) on the EBOV PV surface, adhesion force mapping can be conducted between an AFM cantilever tip treated with Annexin V (A5) and a mica substrate saturated with PV particles. As A5 is a known antibody of PS, this method demonstrates the presence of PS on the particle surface by showing the tip's adhesion to the particle. At the time of abstract submission, this work has not yet been completed, but this will be carried out imminently. Additionally, the method has been proven to work with EBOV-VLP samples.

2.3. UV-Inactivated Ebolavirus

Of the three particles of interest being examined, the sample which is assumed to be most like live Ebolavirus is the UV-inactivated Ebolavirus particles. The only difference between these particles and live EBOV is that the replicative proteins at the nucleus of the virus have been irradiated with UV light, such that the particle is not capable of self-propagating. AFM testing has yet to be conducted on this approximating particle, but from literature information it is expected that UV-EBOV has a long, filamentous structure roughly 50 nm wide and high, and approximately 1000 nm long, and an elastic modulus of the particles of approximately 4-6 MPa. Work to confirm this will be conducted over the next several weeks.

3. Conclusions

The characterization of ebolavirus-approximating particles is critical to the ongoing development of therapeutic drugs to combat the deadly EBOV disease. Overall, VLP demonstrated a physical configuration and elasticity more aligned with the filamentous shape of EBOV and the known stiffness of other Marburg viruses, while PV had a rounder, softer configuration. Additionally, each approximating particle has EBOV GP and PS expressed on the surface. This combined with the fact that VLP is available to the consumer, while PV is not, makes VLP the more attractive approximation for EBOV when conducting experiments regarding therapeutic drug development.

Figure 1

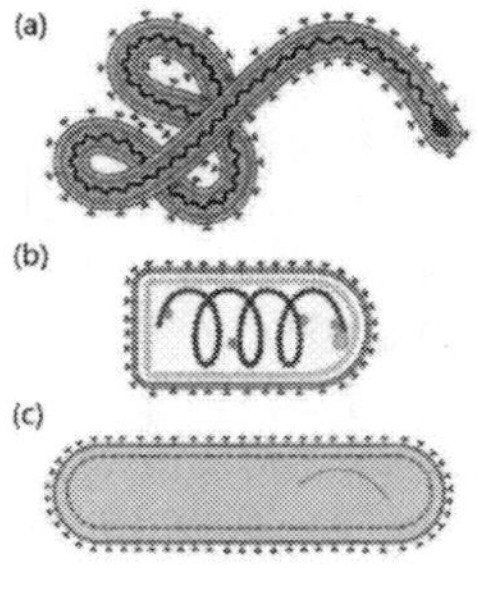

Figure 1. Schematic representation of various EBOV approximations: (a) UV-EBOV, (b) EBOV PV, (c) EBOV VLP. Diagrams produced with Biorender.

Figure 2

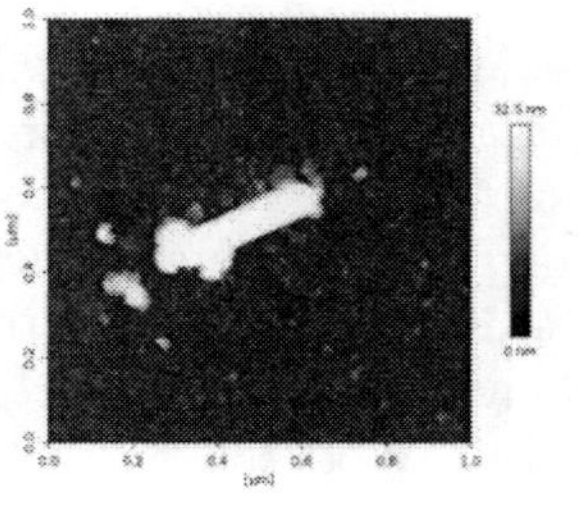

Figure 2. Sample peakforce image of VLP particle adhered to mica, imaged under a PBS droplet.

Figure 3

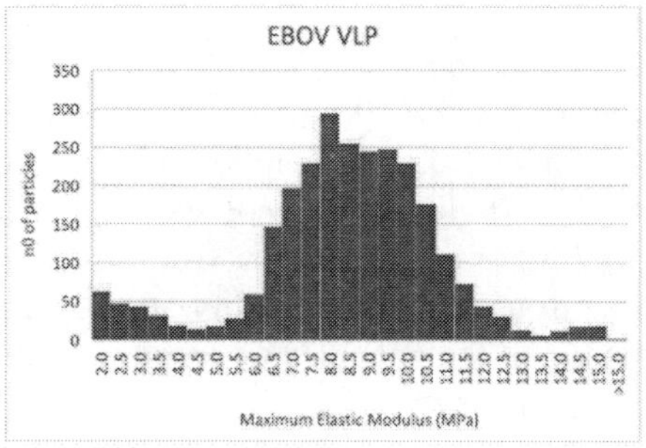

Figure 3. Histogram of the elastic modulus measurements obtained from the VLP nanoindentation force curves.

Figure 4

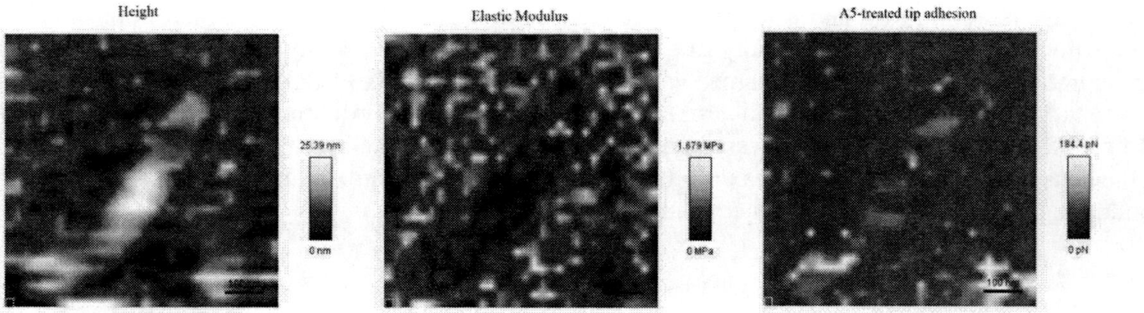

Figure 4. Multiple AFM channels demonstrating the adhesion force between an A5-treated AFM tip and a single VLP adhered to a mica substrate. The first image is the height channel, the second is the elastic modulus channel, and the third is the A5 adhesion channel.

Figure 5

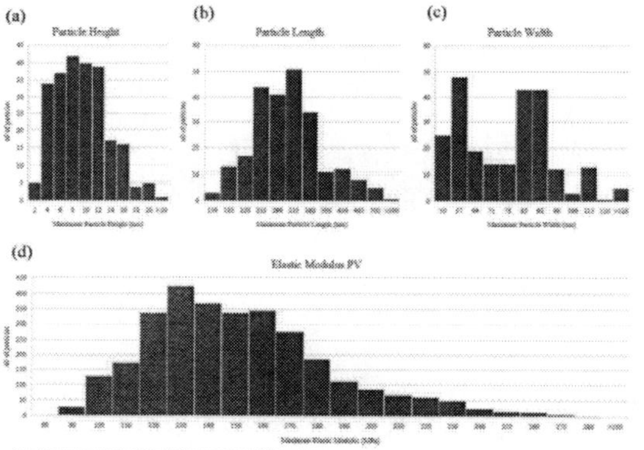

Figure 5. Histogram representation of characterization measurements of the Ebola pseudovirus (PV). (a) Height of the particle, (b) Diameter of the particle (c) Elastic modulus of the PV particle. (d) Elastic modulus of the particle.

183f Silver Nanoparticle Sensor Array for the Detection of Sars-Cov-2 for NSEF Graduate Student Poster Competition

Benjamin Lam

We report a silver nanoparticle (AgNP) sensor array to detect SARS-CoV-2 viral particles utilizing five peptide receptors to produce a colorimetric response. We show that the interaction of the virus particles with the peptides interferes with the shift in plasmonic absorbance and hinders the formation of snowflake-like fractal assemblies between the AgNPs and the bridging peptides. The limit of detection in water of the sensor array was 500,000 viral copies/mL. We demonstrate the capabilities of the sensor array to distinguish between SARS-CoV-2 (Beta, Delta, and Omicron variants) from the influenza virus at the 99% confidence level through linear discriminant analysis. The sensor array was also able to discriminate three out of five negative exhaled breath condensate samples from five positive SARS-CoV-2 EBC samples.

183g Enhancing Capillary Evaporation for Advanced Heat Transfer Applications with PbTe Nanoparticle Coating Layer

Moonchul Park, Tenzin Dorjee, Tahmid Hasan Rupam, Roberto Nunez, Hongbin Ma, Omar T. Ibrahim, Jaewon Lee

In this study, we explore the use of lead telluride (PbTe) nanoparticles as a coating on copper sheets to enhance capillary evaporation for thermal management. PbTe nanoparticles are known for their excellent thermoelectric properties and potential in waste heat recovery and cooling applications. Capillary evaporation, crucial in heat transfer, depends on the surface properties and structure of the material. By using a dip-coating method, we aim to develop structured surfaces that promote efficient heat transfer through capillary evaporation. The impact of surface treatment with Diethyl zinc (DEZ) and Hydrazine on the evaporation rate and hydrophilicity of various solutions is also investigated to assess the improvement in capillary evaporation.

To analyze evaporation rates, 10 μL of super wetting solutions acetone, ethanol, and n-heptane were dispensed onto various surfaces, including a copper sheet (Copper), a PbTe dip-coated layer on copper (PbTe), a PbTe layer with DEZ treatment (PbTe with DEZ), and a PbTe layer with hydrazine treatment (PbTe with Hydrazine). The average evaporation rates for acetone were 0.273 μL/s on Copper, 0.434 μL/s on PbTe, 0.476 μL/s on PbTe with DEZ, and 0.526 μL/s on PbTe with Hydrazine. For ethanol, the rates were 0.0615 μL/s, 0.148 μL/s, 0.160 μL/s, and 0.190 μL/s, respectively. For n-heptane, they were 0.116 μL/s, 0.128 μL/s, 0.172 μL/s, and 0.217 μL/s, respectively. The evaporation rate of all solutions increased in the order of Copper, PbTe, PbTe with DEZ, and PbTe with Hydrazine. Overall, the evaporation rate compared to Copper increased from 10.3% to 209%.

To investigate the hydrophilicity and hydrophobicity of each Copper, PbTe, PbTe with DEZ, and PbTe with Hydrazine surface, a 0.5 μL droplet of DI water was applied to each surface and the contact angle in the equilibrium state was measured. The contact angle in the equilibrium state was 67° for copper, 95° for PbTe, 91° for PbTe with DEZ, and 70° for PbTe with Hydrazine. Subsequently, the evaporation rate was recorded. For the copper sheet, it declined from an initial rate of 0.707 nL/s to 0.226 nL/s at the conclusion of evaporation, indicating a 68.0% reduction. In contrast, the evaporation rate for PbTe began at 0.658 nL/s and decreased to 0.329 nL/s, resulting in a 46% reduction. For PbTe with DEZ treatment, the rate dropped from 0.609 nL/s to 0.418 nL/s, a 31.4% decrease. The PbTe with Hydrazine surface experienced a decline from 0.773 nL/s to 0.475 nL/s, amounting to a 38.6% reduction. However, due to the higher absolute evaporation rate on this surface, evaporation was completed more swiftly than on the other surfaces.

183h Membrane Degradation in Seawater Electrolysis: Understanding Anion Exchange Layer Decomposition Pathways Under Anodic Electrochemical Conditions

Isabela Rios Amador, Ryan Hannagan, Daniela Marin, Ashton Aleman, Dr. Michaela Burke Stevens, Thomas Jaramillo, Adam Nielander

Water electrolysis is a promising strategy to produce H_2 gas for sustainable fuel and as a precursor feedstock to many chemical processes. Engineering electrolyzers that use anion exchange membranes and operate using impure water sources (i.e. seawater) offers a route to reduce water purification costs and unlock durable, affordable, and sustainable H_2 generation. Specifically, anion exchange membrane water electrolyzers (AEMWE) have been identified as an architecture that has the potential to compete with current carbon-intensive H_2 production methodologies. Seawater electrolysis in AEMs combines the key advantages of using cost-effective integral components (i.e. non-precious metal catalysts), generating high purity H_2, lowering ohmic losses, and utilizing alkaline solid electrolyte to suppress corrosive by-product formation (i.e., free chlorine). A major bottleneck for this technology is ionomer degradation that consequently results in device failure.

Here, we discuss the degradation of PiperION, a commonly used AEM, under anodic conditions and in the presence of ionic impurities (e.g, Na^+, Cl^-). These impurities, representative of those expected in impure water feeds (e.g. seawater) can compete with OER to generate corrosive free chlorine species (i.e., Cl_2, HOCl, OCl^-) at the anode. By investigating the electrochemical performance in the presence of these impurities and correlating them to the chemical decomposition products we gain insight to the decomposition mechanism of PiperION. When applying increasingly positive electrochemical potentials to the AEM, we observed changes in the polymer backbone, catalytic selectivity, and the electrochemically active surface area as a function of both current density and ionic concentration. Using X-ray photoelectron spectroscopy (XPS) and electrochemical mass spectroscopy (EC-MS), we detected the change in selectivity as well as products from ionomer decomposition. These insights allow for a better understanding of the ionomer/catalyst effects that are present in water electrolyzer that impact device lifespan and inform the operating conditions necessary to maintain long-term performance. By understanding the chemical degradation products and mechanism of AEMs, our study supports the development of new, more durable materials, which are crucial for accelerating the large-scale production of sustainable H_2 from seawater.

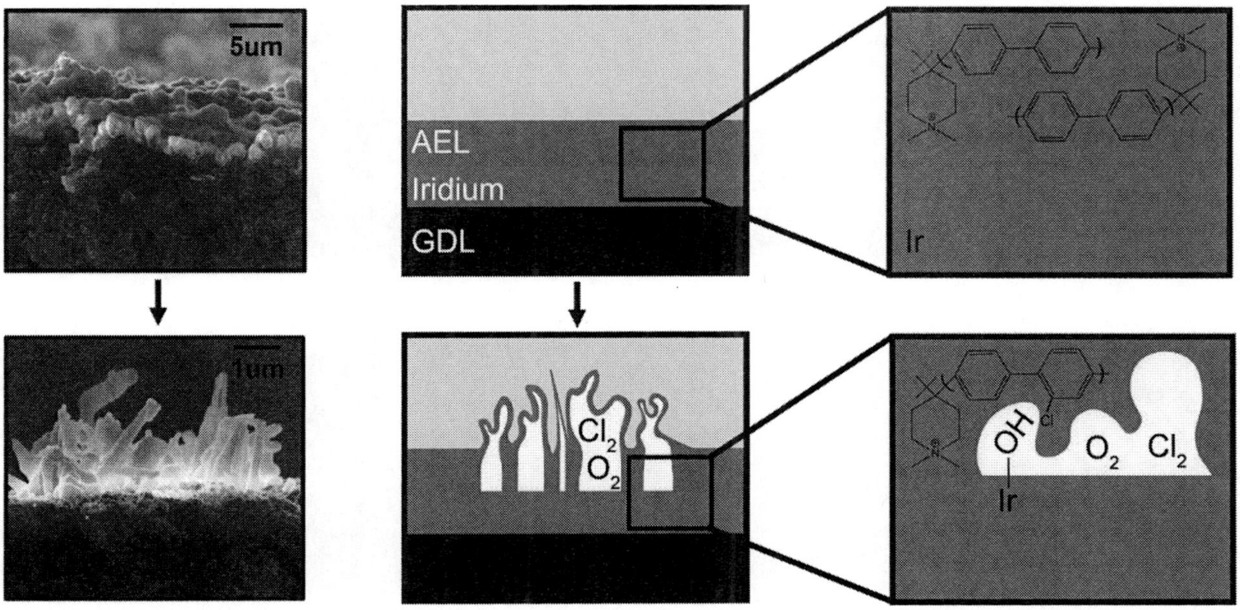

183i Upcycled Polyvinyl Chloride (PVC) Support Layer from Waste PVC Pipe for Thin Film Composite Nanofiltration Membranes

Atta Ur Razzaq, Milad Esfahani

Thin-film-composite (TFC) nanofiltration membranes represent the pinnacle of membrane technology in water treatment and desalination. These membranes typically consist of a polyamide (PA) selective layer and a membrane support layer, often constructed from commercially available materials like polyethersulfone (PES). However, an alternative approach exists that involves using different polymers, preferably upcycled waste polymers, as a viable support layer for TFC membranes. Successfully implementing the upcycling of waste plastics into high-value support membranes can significantly contribute to addressing plastic waste pollution. One of the main challenges associated with utilizing upcycled polymers as support layers is the potential impact on the polyamide selective layer of TFC membranes, subsequently affecting their performance in terms of permeability, rejection, and antifouling properties. This study demonstrates the successful fabrication of TFC membranes with a support layer crafted from upcycled waste PVC pipe. We conducted a comprehensive investigation into the effects of upcycled PVC on the structural and physicochemical properties of the polyamide layer. The ultrafiltration (UF) support membranes were fabricated from waste PVC pipe via non-solvent induced phase separation (NIPS). Subsequently, a polyamide layer was synthesized atop the Upcycled PVC membrane using interfacial polymerization (IP). The physicochemical properties and performance of the TFC membranes with upcycled PVC support layer were compared with membranes with research-grade (RG) PVC and commercial PES support layers. The results unveiled that the TFC membrane with upcycled PVC support layer exhibited higher water permeability (18.2 LMH/bar), in contrast to RG TFC (15.5 LMH/bar) and PES TFC membranes (13.7 LMH/bar). Furthermore, the salt rejection capabilities of the upcycled PVC TFC membrane were competitive and well within an acceptable range.

183j Layer-By-Layer Nanoparticle Delivery of Synergistic Nucleic Acid Therapy for Ovarian Cancer

Eva Cai, Victoria Gomerdinger, Ivan Susin Pires, Margaret Billingsley, Justin A. Kaskow, Paula Hammond, Mae Pryor

Advanced high-grade serous ovarian cancer (HGSOC) is the leading cause of gynecological cancer death in the developed world, with 5-year survival rates of only 25–30% due to late-stage diagnosis and the limitations of current chemotherapies. The primary standard of care consists of tumor removal surgery followed by platinum-based chemotherapy. However, patients often develop resistance to chemotherapies, leading to shortened disease-free intervals, increased risk of metastasis, and worsened prognosis. In particular, the metastatic distribution of HGSOC makes it challenging to reach tumors via systemic drug delivery.

Given that many phenotypes of ovarian cancer (OC) are caused by underlying genetic mutations, small interfering ribonucleic acid (siRNA) therapies have emerged as promising solutions for OC by targeting the underlying mutations with high specificity and high efficiency, with low toxicity compared to systemic chemotherapy. However, bare siRNA is unstable and immunogenic and thus requires a well-engineered delivery system. Layer-by-layer nanoparticles (LbL NPs) are a drug delivery platform in which a charged NP core is layered with alternating charged polymers via electrostatic attraction, enabling the incorporation of complex polymers without the need for complex chemical conjugation techniques. LbL NPs enable broad therapeutic opportunities: by incorporating functional polymers, LbL NPs better avoid liver clearance than existing NPs and selectively target tumors without affecting healthy cells,[2] which in turn reduces drug resistance, improves treatment efficacy, and decreases systemic toxicity. Thus, LbL NPs delivering siRNA therapies targeting multiple tumor proliferation or chemotherapy resistance pathways are a promising solution to improving patient outcomes. In this work, we screened and discovered synergistic siRNA combinations targeting multiple tumor proliferation and chemotherapy resistance pathways, providing a novel therapeutic opportunity.

Previous research demonstrates that LbL NP surface chemistry plays a critical role in governing pharmacokinetics, with particular surface chemistries demonstrating improved tumor targeting abilities, thereby decreasing toxicity associated with systemic drug administration. However, the mechanisms of how surface chemistry affects function and uptake in cancer cells have yet to be elucidated. By modulating the surface polymers, we further optimized the LbL NPs to selectively traffick siRNAs to tumors while reducing off-target effects to healthy cells. By engineering LbL NPs to deliver combination siRNA therapies, we show improved treatment efficacy and decreased systemic toxicity in an in vivo murine model of HGSOC, demonstrating a promising solution to improving patient outcomes.

183k Influence of Surface Parameters and Encapsulant Type on Nano-Pesticide Adsorption Thermodynamics on Plant Cuticle

Yashwanth Arcot, Minchen Mu, Luis Cisneros-Zevallos, Mustafa Akbulut

Foliar application of pesticides is a widespread practice due to the rapid uptake of active ingredients by leaves through stomata, leaf cuticles, and trichomes, and the absorbed compounds are transported to other plant tissues via the symplast and apoplast pathways. Nano-pesticides represent the most effective form of pesticides for foliar administration, as they enhance solubility in the aqueous phase, improve bioavailability, and achieve sustained release of active ingredients. To optimize the efficacy of nano pesticides and enhance their adhesion to the plant cuticle and epicuticular waxes during spraying, it is crucial to comprehensively investigate the structural, surface, and rheological parameters that influence the thermodynamics of their adhesion.

To investigate the impact of hydrophobic, electrostatic, hydrogen bonding, steric stabilization, and Lewis's acid-base interactive forces on the stability of the nano-pesticides and adsorption of nano-pesticides on epicuticular waxes, various nanocarriers were utilized including pluronic F-127, a triblock copolymer encapsulant; whey protein isolate, a natural protein encapsulant; Tween 80, a non-ionic surfactant; ethyl lauroyl arginate, a cationic surfactant; sodium dodecyl sulfate, an anionic surfactant; and Cocamidopropyl betaine, a zwitterionic surfactant. The nanopesticide systems were designed to maintain consistent concentrations of the active ingredient and uniform particle sizes, allowing for a focused examination of the effect of interactive forces.

Further, to elucidate the influence of surface roughness on nanopesticide adhesion, the findings from adsorption studies conducted on the adaxial side of leaf surfaces were compared with those obtained from studies on substrates coated with waxes mimicking the composition of epicuticular wax found on leaf surfaces, but lacking surface roughness.

The nano-pesticide systems, substrates, and their interactions were characterized using dynamic light scattering (DLS), scanning electron microscopy (SEM), atomic force microscopy (AFM), contact angle analysis, UV-visible spectroscopy, and fluorescence microscopy. Additionally, the kinetics of adsorption and the fate of the nano-pesticides on the wax substrates were demonstrated by monitoring changes in the frequency (Δf) and dissipation (ΔD) of substrate vibrations using quartz crystal microbalance with dissipation (QCM-D), which signifies the type of nanopesticide and substrate interactions.

Based on the results, we concluded that electrostatic forces of interaction and steric repulsion between the nanopesticide system and the substrate, along with topographic roughness, are the most dominant factors influencing the adsorption of nanopesticides on the leaf cuticle and wax.

1831 Award Submission: *Quantifying Enzymatic Small Extracellular Vesicles (sEVs) for Metastatic Cancer Staging Via Magnetic Nanoporous Membrane (MNM) Activity Assay*

Tiger Shi, Chenguang Zhang, Youwen Zhang, Xuemin Lu, Gaeun Kim, Sonu Kumar, Ceming Wang, Yichun Wang, Satyajyoti Senapati, Hsueh-Chia Chang

Cancer metastasis is the leading cause of death for cancer patients. Active surface proteins like membrane-type matrix metalloproteinases (MT-MMPs) and a-disintegrin and metalloproteinases (ADAMs) have been identified to be highly expressed in later stages of disease. Additionally, increased metalloproteinase (MP) expression has been observed prior to tumor metastasis to remodel distal extracellular matrices (ECM) readying the site for invasion and proliferation. Small extracellular vesicles (sEVs) – ranging from 20-200 nm – have been shown to display parent cell membrane enzymes on their lipid bilayer surface. Affinity assays offer little metastatic information captured by the activity level of the enzyme. Alternatively, existing bulk plasma activity assays for these enzymes lack sensitivity and are prone to interfering agents found in plasma.

In this work, we overcome these limitations with a Magnetic Nanoporous Membrane (MNM) diagnostic technology which uses superparamagnetic beads to rapidly isolate and enrich specific tumor cell-originated sEVs. A rapid sample incubation (30-minutes) is performed using 0.22-um filtered plasma with antibody-functionalized nanobeads. The bound sEVs are trapped by the strong field gradient present at the sharp edges of the MNM pore edge at >99% capture. A controlled wash is applied to remove non-specifically bound entities allowing for direct characterization from minimally pretreated blood plasma. The activity assay utilizes molecular beacons synthesized from enzyme-specific substrates that fluoresce upon enzymatic digestion. The platform is capable of running up to 12 wells simultaneously and can be read-out on a custom-made 96-well plate using a standard laboratory fluorescence plate reader. This assay has an overall turn-around time of two to three hours.

We establish the localization of the majority of ADAM10 enzymatic activity from cancer cells to the surface of sEVs. ADAM10+ sEV activity normalized to sEV concentration is indicative of metastatic potential in prostate cancer cell lines. Interestingly, we find that lysed sEV sample produces significantly less activity signal, suggesting that the activity of ADAM10 is conferred by its specific conformation on the sEV lipid bilayer. ADAM10+ sEV activity hence produces significant separation as a general metastatic staging biomarker in mCRC patients. The findings recapitulate existing understandings of ADAM10 in cancer metastasis and focus the field towards the sEV-bound partition. Work on the MNM well-plate platform is ongoing with significant promise in improving sensitivity down by one to two orders of magnitude compared to the bulk assay (sub-femtomolar). Our group is currently working on selective pin-down of cancer-specific EGFR+ sEVs in a larger clinical cohort of mCRC patients.

415a Development and Early Testing of a Prototype Device for Nanosensor-Enabled Chemical Cystoscopy

Wonjun Yim, Xiaojia Jin, Daniel A. Wollin, Mark Preston, Michael S. Strano

Bladder cancer is a very common and morbid malignancy. Owing to the high recurrence rate and relatively slow progression, patients require frequent surveillance and treatment, making bladder cancer the most expensive malignancy per person. Cystoscopy and biopsy are considered the gold standard for diagnosis, but, although blue-light cystoscopy and other visualization technologies may improve detection, office cystoscopy may often lead to a delay in treatment or a misdiagnosis. Our goal is to develop a novel sensing modality for biomarker detection of bladder cancer during office cystoscopy to improve accuracy.

We have developed single-walled carbon nanotubes engineered with specific polymers in which analytes binding events induce fluorescent changes for biomarker detection. This technique is called Corona-Phase Molecular Recognition (CoPhMoRe). Using CoPhMoRe, we aim to detect biomarkers related to bladder cancer. The nanosensors with promising specificity/sensitivity are coated on a medical catheter to produce a 3D chemical map. The nanosensor readout is performed through an optical fiber within the catheter. Nanosensor-enabled chemical cystoscopy (NECC) is composed of the nanosensor catheter and optical fiber passed through the channel of a cystoscope to allow for in-office biomarker assessment.

We successfully coated nanosensors targeting H_2O_2 on a catheter; the sensors show unique fluorescent peaks from 900 to 1200 nm. We validated NECC's ability to capture biomarker release using T-24 bladder cancer cells, which were cultured on glass slides and analyzed with the NECC catheter in solution. Upon benzidine stimulation, a sharp decrease in nanosensor intensity was observed at the bottom of the catheter, which corresponds to the presence of H_2O_2 release near the cells, while there was no change in intensity without stimulation. In this study, we constructed a prototype of a nanosensor-coated catheter for eventual bladder cancer detection using an optical fiber to capture nanosensor fluorescence output. The system was able to assess proximity to T-24 bladder cancer cells due to H_2O_2 concentration gradients.

415b Computational Analysis of Photophysical Changes of Oligomeric *P*-Phenylene Ethynylene upon Binding to Amyloid-β Aggregates

Benjamin Fetrow, David G. Whitten, William P. Bricker, Eva Chi

Background and Motivation

Alzheimer's disease is an uncurable progressive neurodegenerative disease and is the most common cause of dementia. As current understanding of the disease suggests damage begins accumulating a decade before diagnosable symptoms, preventative treatment strategies will require screenings in absence of symptoms. The high cost of PET and MRI scans make them unsuitable for the throughput necessary to screen the large population of 65+ individuals most at risk of developing Alzheimer's disease. An alternative method is near-IR fluorescence imaging, with a lower cost and a less invasive procedure. We have reported a small-molecule fluorescent sensor able to selectively detect and oxidize the amyloid-β oligomers implied as pathogenic agents in the early development of Alzheimer's disease. In this study, we use computational modeling to gain a greater understanding of what changes in molecule-protein binding lead to both turn-on fluorescence and turn-on singlet oxygen generation.

Methodology

We utilize all-atom molecular dynamics to model fluorophore molecule behavior in environments of interest. These environments being a singular monomer, a dimer, and a dimer associated with the surface of an amyloid-β aggregate. We first extract dihedral angles associated with the conjugated pi-system backbone to measure the overall planarity of the fluorophores. We also periodically extract snapshots of molecular conformation upon which we perform density functional theory and time-dependent density functional theory *ab initio* calculations. These calculations are used to obtain vertical excitation energies and oscillator strengths for said transitions, from which we can gain an understanding over the absorbance behavior of the fluorophores in these conditions. We further use transition density cube calculations to compute the impact of Coulombic coupling interactions upon dimer excitation energies. These computations are compared against experimental spectra.

Results and Conclusions

The planarity of the fluorophore's chromatophore backbone is controlled by rotation around two triple bonds, in turn forming two dihedral angles. For the monomer, the full possibility space of these angles is inhabited in a random fashion. Upon dimerization, we find the molecules are restricted to only planar conformations while still being able to transition between these conformations. When further restricted by adherence to the protein aggregate surface, the molecules retain the restriction to planar conformations, but are largely restricted from transitioning between different planar states.

The computed vertical excitations for the monomer reflect the random conformations. We define excitation to the S1 state by the lowest calculated transition with appropriate energy and non-negligible oscillator strength. For the monomer, this transition is not always available, with the lowest energy available transition being the experimental S2 transition for many out-of-plane molecular conformations. The S1 transition stabilizes upon dimerization, always being accessible and typically being the lowest energy available transition calculated. This reflects the condition's tendency for planarity, as the molecules spend greater time in the planar conformations where the S1 transition is favored. It is yet further stabilized upon protein adherence, where for all snapshots the lowest energy computed excitation has a non-negligible oscillator and so corresponds with the S1 transition. By combining all calculated vertical transitions into a spectrum, and then combining all snapshot spectra into a single spectrum, we can create computational spectra corresponding to the bulk steady-state conditions of experimental spectra. We see a redshift and a strengthening of the S1 transition upon dimerization, and a further redshift upon protein association.

These calculations only describe how geometry changes upon dimerization alter fluorophore behavior. To expand into dimer interactions, we use the transition density cube method to calculate Coulombic interaction energies. This method assumes dimers are Kasha-like: possessing strong coupling, but without any meaningful charge-transfer states and without substantial vibronic coupling. The dimers alone show largely H-aggregate behavior, with reasonably strong transition dipoles and positive interaction energies. The dimers switch to largely J-aggregates upon association with protein, possessing negative interaction energies. These calculations suggest further redshifts and stronger absorbance upon protein adhesion.

All the computational data point to a redshift and increased S1 absorbance upon dimerization, with both strengthening upon association with protein aggregates. First through conformational changes induced by rotational restriction upon dimerization, and then through a transition from H- to J-aggregates. This behavior is reflected experimentally. The monomer form is highly disfavored in the aqueous conditions of interest for potential clinical use. However, the computed redshift and strengthened absorbance is seen experimentally upon addition of protein aggregates.

This study investigates two potential sources of increased absorbance intensities upon protein binding. Increased absorbance intensity does correlate with increased fluorescence intensity, and the conversion from H- to J-aggregates provides another suggestion for the turn-on fluorescence observed. However, future work directly analyzing emission behavior is required to gain a clear understanding of the source of observed fluorescence turn-on.

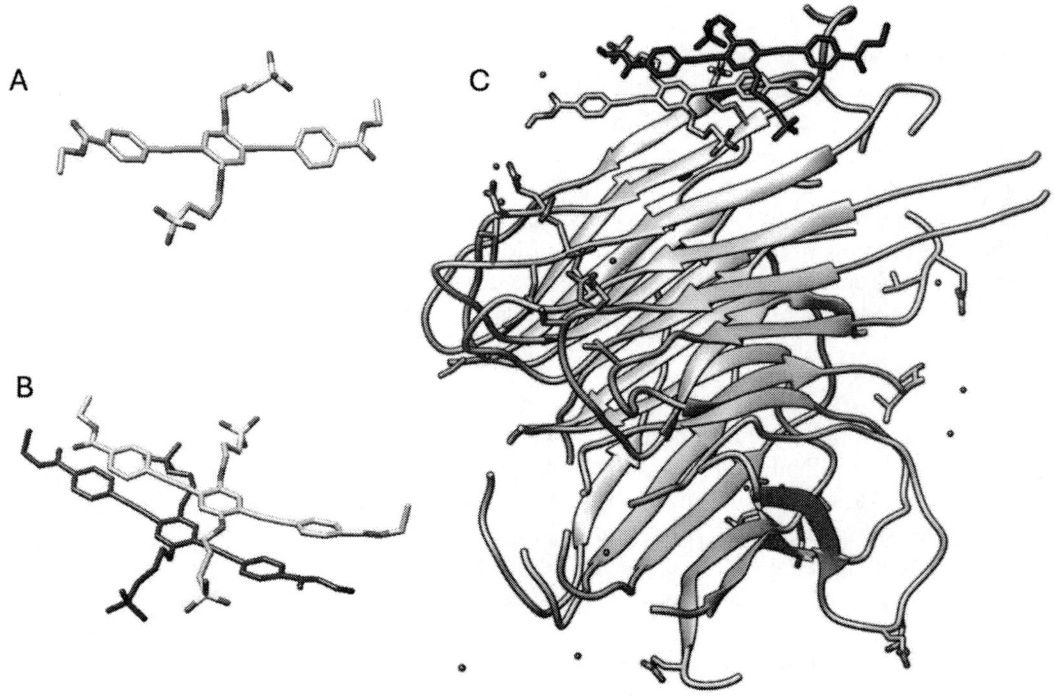

Figure 1: Representative snapshots of an OPE monomer in solution (A), an OPE dimer in solution (B), and an OPE dimer bound to an Aβ40 oligomer (PDB ID 2LMN) (C).

415c Imaging Dopamine Signaling Aberrations in Cocaine Sensitized Animals with Near-Infrared Fluorescent Nanosensors

Natsumi Komatsu, Nicholas Ouassil, Markita Landry

Cocaine locomotor sensitization is a well-established model for understanding the physiological effects of cocaine abuse. Previous studies have established a loss of dopamine D2 autoreceptors in striatal tissue as a consequence of locomotor sensitization, yet how these differences in autoreceptor expression affect dopamine signaling remains unclear. Herein, we utilized non-genetically encoded near-infrared fluorescent nanosensors based on carbon nanotubes to image dopamine modulation in acute brain slices. By leveraging dopamine D2 receptor agonists along with the dopamine nanosensor, we developed a method to identify changes in the density of functional D2 autoreceptors and dopamine signaling aberrations that accompany sensitization, at the level of individual dopamine synapses. We apply our method to quantify changes in D2 autoreceptor activity in the nucleus accumbens core in mice after prolonged cocaine exposure. Our preliminary data indicates a decrease in the number of D2 autoreceptors in cocaine sensitized mice, consistent with prior studies, and suggests that cocaine abuse leads to higher levels of dopamine signaling relative to non-cocaine treated mice. Furthermore, we confirmed that dopamine signaling changes by application of D2 receptor agonists are from loss of individual dopamine release sites rather than changes in the quantity of dopamine released, or changes in dopamine reuptake kinetics. Our study provides a route to understand the changes in neuromodulation that underlie the neurobiology of substance abuse.

415d Nano-Patterning of Multiplexed Bioreceptors Using Thermal Scanning Probe Lithography

Hashem Nasralla, Alexander Wright, Rahul Deshmukh, Davood Shahrjerdi, Elisa Riedo

Achieving high-resolution multiplexed chemical nanopatterns on surfaces is crucial for biomedical applications, including lab-on-a-chip systems, biosensing, tissue engineering, and cell manipulation studies.

However, technology facilitating the precise nano-assembly of different biomolecules, such as antibodies or aptamers, at specified locations on a surface remains elusive. This challenge stems from various factors, including resolution limitations, the diversity of biomolecule types, spatial accuracy, and the risk of non-specific binding of analytes to unintended locations. Thermal Scanning Probe Lithography (t-SPL), a nanofabrication technique that involves using a heated silicon probe to locally modify material on a substrate, addresses these issues by selectively patterning thin film surfaces with nanoscale precision to locally activate chemical reactions in an active polymer or resist layer, or directly remove material to expose chemically activated layers underneath a protective resist [1, 2].

To that end, we present an adaptable process, implementing thermal Scanning Probe Lithography, to conjugate different biomolecular receptors, including antibodies and aptamers, to targeted patterns with sub-20 nanometer resolution and a minimum pitch of 200 nanometers [3].

In our study, thin polymer films are deposited on silicon wafers by spin coating. Using t-SPL patterning, the polymer is heated to locally deprotect and expose amine groups for desired surface functionalization, including biotin-streptavidin interactions, thiols, or click-chemistry. Consecutive polymer patterning and surface functionalization steps by drop-casting with different bioreceptors create high resolution patterns with the ability to independently detect a variety of target molecules. Furthermore, *in-situ* t-SPL topographical imaging and atomic force microscopy (AFM) are used to verify the presence of each bioreceptor after each "patterning and functionalization" step by the changing in pattern depths. This approach enables the creation of chemically or biologically active surfaces for biomedical applications without the need for physical macroscopic barriers that limit the scalability of a system.

1. Albisetti, E., et al., *Thermal scanning probe lithography.* Nature Reviews Methods Primers, 2022. **2**(1).
2. Liu, X.Y., et al., *Sub-10 nm Resolution Patterning of Pockets for Enzyme Immobilization with Independent Density and Quasi-3D Topography Control.* Acs Applied Materials & Interfaces, 2019. **11**(44): p. 41780-41790.
3. Wright, A., et al. *Transistors platform for rapid and parallel detection of multiple pathogens by nanoscale-localized multiplexed biological activation (Under Review).* 2023.

415e Bioconjugated Magnetic Nanoparticles for Non-Invasive Diagnosis of Inflammatory Bowel Disease

Shno Asad, David Ahl, Mia Phillipson, Alexandra Teleki

Superparamagnetic iron oxide nanoparticles (SPIONs) have emerged as versatile tools in biomedical research, promising innovative solutions for disease treatment, diagnosis, and monitoring. SPIONs are suitable as contrast agents for magnetic resonance imaging (MRI). However, their potential for targeted diagnosis of gastrointestinal (GI) diseases such as inflammatory bowel disease (IBD) has so far been limited. In this study, we have investigated the targeting capability of bioconjugated SPIONs towards inflamed intestinal tissue and developed a protocol for their in vivo administration and subsequent ex vivo evaluation using imaging techniques. We synthesize and successively modify silica-coated SPIONs with organic moieties to enable click chemistry-based bioconjugation with anti-ICAM1 and anti-CEACAM1 antibodies. These antibodies target proteins that are upregulated on the epithelial cell surface during inflammation in the GI tract.[1] Thermogravimetric analysis and Fourier transform infrared spectroscopy confirm successful functionalization of SPIONs following each modification step. Assessment of cell viability indicates no adverse toxic effects of the bioconjugated particles.

Our functionalization approach facilitates the controlled orientation of the ligand on the particle surface, exposing the antigen-binding site to enable particle binding to target sites. Quantitative analysis reveals a significantly higher concentration of iron in inflamed Caco-2 cells upon exposure to ICAM1-modified particles compared to non-conjugated counterparts. Furthermore, laser scanning confocal microscopy of inflammation-induced Caco-2 cells suggests surface interaction and internalization of the bioconjugated SPIONs, highlighting their potential for targeted imaging in inflammatory diseases. Further, we designed an experimental protocol for incubating bioconjugated nanoparticles in the colon of mice in vivo. Inflammation was chemically induced in the GI tract of the mice prior to nanoparticle administration to resemble ulcerative colitis in humans. Our findings suggest that rectal administration can be an effective approach for nanoparticle delivery to the colon. The developed experimental procedure enabled exploration of the targeting potential of the bioconjugated SPIONs using laser scanning confocal microscopy and MRI. Confocal micrographs showed preferential accumulation of Ceacam1-conjugated SPIONs in severely inflamed regions, while particles in moderately damaged areas were primarily localized above the epithelial layer, hindered by the presence of the mucus barrier. This spatial distribution indicates the utility of targeted nanoparticles for characterizing the degree and severity of the active inflammation. Overall, our findings demonstrate the potential of SPION-based systems for precise imaging and therapeutic interventions in GI diseases.

Acknowledgement:

This project has received funding from the European Research Council (ERC) under the European Union's Horizon 2020 research and innovation programme (grant agreement No. 101002582). The authors also acknowledge financial support from the Science for Life Laboratory.

References:

1. Asad, S., C. Wegler, D. Ahl, C.A.S. Bergström, M. Phillipson, P. Artursson, and A. Teleki, "Proteomics-informed identification of luminal targets for *in situ* diagnosis of inflammatory bowel disease," **J. Pharm. Sci.** *110*, 239 (2021).

415f Photoacoustic Image-Guided Magnetic Delivery and Retention of Dual Nanoengineered Stem Cells

Anamik Jhunjhunwala, Kelsey P. Kubelick, Masako Ishikawa, Brooke Chambliss, Myeongsoo Kim, Jinhwan Kim, Stanislav Y. Emelianov

Over the past two decades, stem cell therapy (SCT) has garnered significant attention as a promising strategy for tissue repair/regeneration and functional restoration. Despite encouraging outcomes in numerous preclinical trials for treating spinal cord injuries, osteoporosis, and cardiac pathologies, translating SCT to clinical settings remains hampered by significant challenges. A major limitation lies in the poor delivery efficiency and inadequate long-term retention of stem cells at the target site. Minimally invasive delivery routes, such as intravenous (IV) administration, are preferred but often result in suboptimal engraftment and delocalization of the transplanted cells from the injury site over time. Additionally, locally administered stem cells exhibited poor long-term retention at injury sites. Regardless of cell type or delivery route, acute retention of less than 10% is generally reported. Therefore, novel approaches are required to deliver transplanted stem cells to their intended target regions and facilitate their sustained engagement with the damaged tissue.

Magnetic targeting is a common technique used to address the challenge of poor stem cell retention. This technique is based on manufacturing of magnetically-targeted stem cells followed by *in vivo* targeting with the aid of magnetic fields. While other researchers have employed magnetic guidance strategies with encouraging outcomes, their evaluation has primarily relied on end-point analysis methods, such as histology, which fail to reflect the real-time dynamics of the tissue region. Magnetic resonance imaging (MRI) has been a commonly explored modality for tracking transplanted cells, but it does not allow for image-guided delivery or real-time monitoring of their spatial distribution. Additionally, the strong magnetic fields inherent to MRI can interfere with maintaining magnetically-targeted stem cells at the treatment site. Hence, there is a critical need for a non-invasive imaging approach that allows for real-time monitoring of magnetically-guided stem cell delivery and distribution within the target tissue region, without disrupting the magnetic targeting process itself.

Our approach decouples the targeting mechanism from the imaging modality by employing an external magnetic field to guide magnetically-labeled stem cells and photoacoustic imaging to monitor the delivery/distribution. Silica-coated iron oxide nanospheres (IONPs) facilitate magnetic cell retention and delivery, while silica-coated gold nanorods (AuNRs) enable ultrasound/photoacoustic (US/PA) image-guidance and monitoring **(A)**. US imaging provides excellent anatomical images of tissue with high spatial resolution, while complementary PA imaging can convey functional information with high contrast, resolution, imaging depth, and sensitivity. Successful synthesis and morphology of the dual nanoparticles were confirmed through transmission electron microscopy (TEM), revealing well-defined nanoparticles with distinct silica shells **(B)**. UV-vis-NIR spectroscopy and dynamic light scattering were employed to assess optical absorbance and surface charge. Photostability of AuNRs was evaluated under pulsed laser irradiation, while magnetic retention capability of IONPs under simulated venous flow conditions was assessed.

Following synthesis and characterization, human adipose tissue-derived mesenchymal stem cells (MSCs) were labeled with the dual-nanoparticle system. *In vitro* photoacoustic (PA) imaging (20 MHz, 680–970 nm, Vevo/LAZR, Visual Sonics Inc.) using a tissue-mimicking phantom was used to characterize the PA signals from dual-labeled MSCs. The photoacoustic spectrum of labeled MSCs exhibited spectral characteristics consistent with AuNRs, and US/PA imaging of cells at decreasing concentrations revealed a detection limit of 10-20 cells/μL. **(C, E)**. Based on cytotoxicity studies and demonstrated US/PA sensitivity, the optimal labeling concentration was determined. Magnetic homing capability was demonstrated using a cell-laden tissue-mimicking phantom and an external magnet (1 cm distance, 5 min exposure), resulting in a 2.4-fold increase in PA signal intensity on the side nearest the magnetic field, indicating enhanced cell trafficking and localization **(D, F)**. The control condition showed no signal difference, confirming the specificity of magnetic homing facilitated by the dual-nanoparticle system. **(D,**

F).

To demonstrate photoacoustic image-guided stem cell delivery in vivo, mice received intravenous injections of dual-labeled MSCs via the tail vein (100 µL, 500 cells/µL) for US/PA imaging (40 MHz, 680–970 nm, Vevo/LAZR, Visual Sonics Inc.). A magnet was placed over the femoral artery to demonstrate magnetic trapping at the hind limb. Multiwavelength photoacoustic datasets were acquired before injection, immediately following 30 minutes of exposure to visualize magnetic trapping of dual-labeled MSCs, and 30 minutes after removing the magnet to visualize dual-labeled MSC clearance. Control groups consist of mice that received dual-labeled MSCs with no magnet, and mice that received no cell injection. In the magnetic retention group, a higher concentration of labeled stem cells localized in the hindlimb region, demonstrating successful magnetic targeting and delivery **(G)**. In contrast, control mice exhibited dispersed stem cells in vasculature with no preferential localization **(G)**.

Our findings demonstrate the effective magnetic guidance and retention of dual-labeled MSCs to target regions, facilitated by an external magnetic field and monitored non-invasively in real-time through photoacoustic imaging. Unlike previous approaches that either perform magnetic delivery without visualization or real-time guidance or employ MRI-assisted visualization without strategies for guidance or retention, our multifaceted approach synergistically permits for photoacoustic-guided magnetic delivery of stem cells allowing us to localize cells at site-specific regions within the body. The imaging feedback mechanism enables dynamic adjustments to the administration injection location, optimizing stem cell delivery efficacy. This innovative approach promises to advance clinical stem cell therapy by enhancing cell retention and enabling precise, noninvasive tracking within the body.

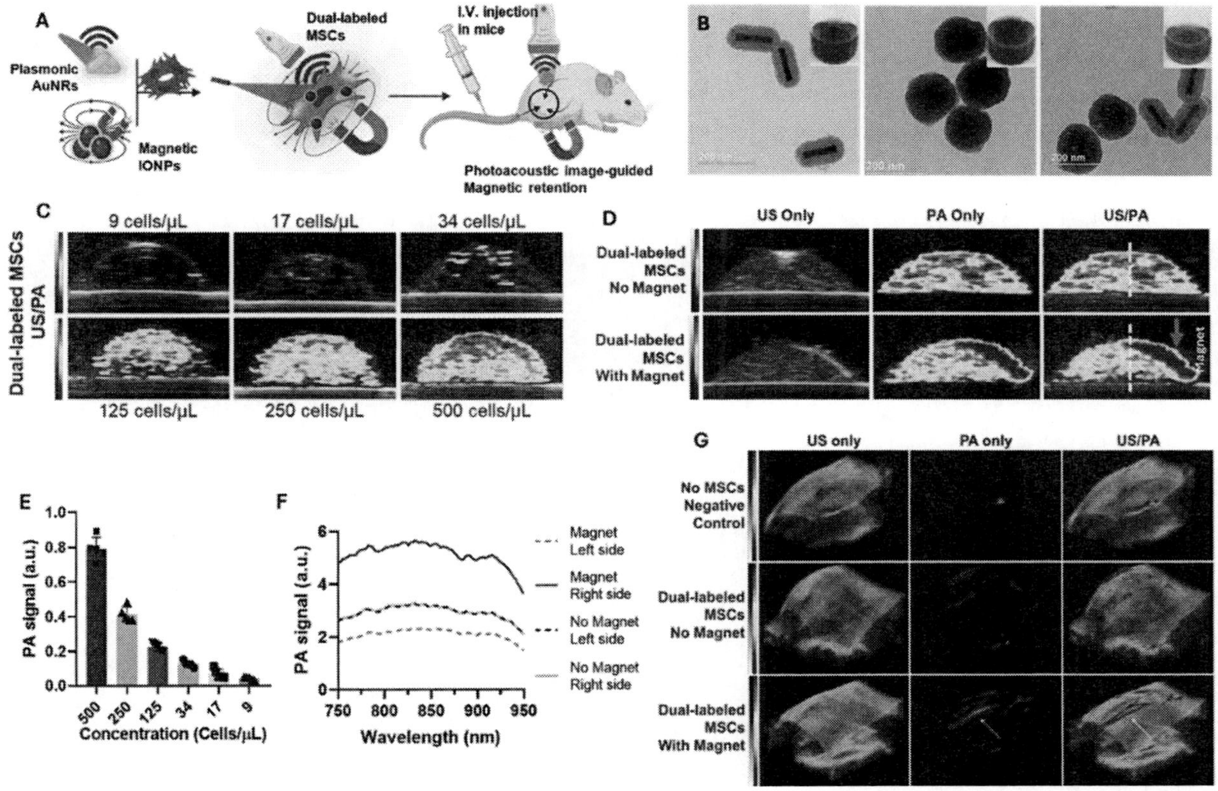

415g Design Strategies for Countering the Effect of Fluorophore-Quencher Labelling on DNA Hairpin Thermodynamics

Lin Yue Yung

In this work, we studied the effect of fluorophore modifications on DNA hybridization thermodynamics with a focus on hairpin opening. We observed an extremely low hybridization efficiency of single stranded trigger strand with end-modified hairpins. The hybridization efficiency improved drastically, i.e. up to ca. 40-folds, by incorporating an external hairpin stem structure on the single stranded trigger strand to promote coaxial stacking hybridization. An alternative counter strategy was to modify the dye internally in the hairpin stem to destabilize its closed state. Lastly, the effect of dye modifications on hairpin opening was similarly observed in the more complex molecular operation of Hybridization Chain Reaction (HCR). Overall, the study affirmed that the careful choice of fluorophore-quencher pair, aside from their spectral properties, and their strategic placement should be accounted for in the design of oligo probes which is particularly crucial for probes with hairpin structure.

414a Surface-Capped Protein Nanoparticles for Non-Viral Gene Delivery

Fjorela Xhyliu, Yao Yao, Joerg Lahann

Gene therapy holds great promise for addressing critical genetic diseases and cancers. The modification or correction of disease-causing genetic mutations through the delivery of nucleic acids is of high interest. Successful delivery of nucleic acids such as plasmids (pDNA) and messenger RNA, to the cells requires overcoming certain biological barriers; these include nuclease degradation, short half-life in the bloodstream, and internalization into target cells. Nanoparticles (NPs) have played a significant role in advancing the field of non-viral gene delivery. However, clinical translation remains a challenge for most NP systems. These challenges are predominantly due to poor NP stability, inefficient nucleic acid encapsulation, delivery to the desired cells, and effective transfection within the cells. Protein nanoparticles (PNPs) are a promising platform for nanotherapeutics, offering biological and physical properties suitable for overcoming barriers to clinical translation. This work explores the surface stabilization of PNPs with polycations to improve NP stability and enhance pDNA transfection both *in vitro* and *in vivo*.

Utilizing electrohydrodynamic (EHD) jetting, serum albumin was used to encapsulate a large pDNA. The as-jetted PNPs were collected in an aqueous polycation solution, resulting in the layer-by-layer deposition of positively charged nanoparticles, with the polycation stabilizing the negatively charged protein nanoparticle surface. These nanoparticles are called surface-capped PNPs (scPNPs). Electron microscopy and dynamic light scattering (DLS) confirmed that the scPNPs have sub-200 nm diameters in their dry and hydrodynamic states. Moreover, the scPNP size and positive charge were stable in aqueous conditions for more than 2 weeks. A quantification method utilizing nanoparticle tracking analysis (NTA) and UV-visible absorbance, was established to determine the pDNA content per scPNP, which indicated an overall yield of 75-80%. Cellular uptake kinetics of scPNPs were evaluated in two different cell lines, and 100% cellular uptake was achieved within 6 hours of incubation. In human HepG2 cells, transfection rates of 50% were attained by varying the pDNA amounts per scPNP and scPNP dosing, with negligible cytotoxicity observed. Preliminary results in a mouse glioblastoma model indicate that scPNPs demonstrate higher transfection efficacy than JetPEI®, a commercial transfection agent.

This work aims to advance the fundamental understanding of design choices when optimizing the efficiency of large nucleic acid internalization, delivery, and transfection efficacy. The scPNP platform promises to incorporate a variety of payloads – such as siRNA, mRNA, antisense oligonucleotides, and Cas9-sgRNA complexes – for an expansion into numerous therapeutic applications tackling diverse cancers and genetic disorders.

414c The Green Chemistry Synthesis of Cationic Lipids and Polymers for Nanoparticle-Based mRNA Vaccines and Therapeutics

Xiaoyang Xu

Lipid nanoparticles (LNPs) formulated with ionizable cationic lipids are clinically advanced non-viral gene delivery platforms and hold great potential for gene therapeutics, exemplified by the two COVID-19 vaccines developed by Pfizer-BioNTech and Moderna. The FDA-approved ionizable cationic lipids such as DLin-MC3-DMA (MC3), ALC-0315, and SM-10218 are synthesized through multistep reactions and present drawbacks such as a low chemical yield especially after multiple post-reaction processes. Moreover, most published lipid-like materials are synthesized through Michael addition or epoxide ring-opening reactions, which makes the resulting lipid-like materials nonbiodegradable with potential cytotoxicity. Herein, we propose a 'Green Chemistry' approach to develop a new chemical library of ionizable cationic lipids or polymers using a one-step chemical-biological enzyme-catalyzed esterification method. Subsequently, these lipids or polymers are formulated into LNPs or polymer-lipid hybrid nanoparticles for mRNA delivery and COVID-19 vaccine development. For example, we have utilized a chemical and biological tool, the enzyme catalyst Candida antarctica Lipase B, to synthesize a chemical library of degradable ionizable cationic lipids using amino alcohols and lipid acids, resulting in 144 lipid-like materials with high chemical yield and purity. Through highthroughput screening, the top-performing ionizable lipid was determined to be AA3-DLin, which is rationally designed to be composed of two unsaturated linoleic lipid chains, two tertiary amine head groups from piperazine which can provide enough positive charge to condense the gene payloads in acidic buffer, and two biodegradable ester linkers. To further investigate the optimized AA3-DLin LNP formulation, an orthogonal design of experiment (DoE) methodology was employed to fine-tune the molar ratios of AA3-DLin, DOPE, cholesterol, and DMGPEG. The top-performing AA3-DLin LNPs show outstanding mRNA delivery efficacy and long-term storage capability. Furthermore, the AA3-DLin LNPs efficiently deliver SARS-CoV-2 full-length spike encoded mRNA in vitro and in vivo. Both fresh and 12-month stored AA3-DLin COVID-19 vaccines successfully elicit spike-specific antibodies and Th1-biased T cell immune response in immunized mice. The AA3-DLin LNPs are an excellent mRNA delivery platform, and this study provides an overall perspective of new ionizable cationic lipids from aspects of lipid design, synthesis, screening, optimization, fabrication, characterization, and application.

414d Impact of Surface Chemistry on Lipid Nanoparticle Transfection and Targeted Delivery

Namita Nabar, Tamara Dacoba, Gil Covarrubias, Denisse Romero-Cruz, Deeksha Kumaresh, Paula Hammond

Introduction

Lipid nanoparticles (LNPs) are potent delivery vehicles for multiple nucleic acid cargos, including messenger RNA (mRNA) and plasmid DNA. These systems have been used clinically for liver-targeted delivery of nucleic acid therapeutics, and their high transfection efficacy makes them valuable for treating diseases beyond the liver, such as sickle cell disease and extrahepatic cancers. However, targeted intravenous LNP delivery to extrahepatic spaces is challenged by multiple physiological parameters. NP surfaces adsorb negatively charged serum proteins, which induce aggregation and activate rapid sequestration by immune cells [1]. In particular, LNPs adsorb apolipoprotein E (ApoE), which enhances accumulation in the liver by binding to hepatocytic low-density lipoprotein (LDL) receptors [2]. Moreover, efficient gene delivery must mitigate off-target LNP uptake by more populous cell types. LNP surfaces for targeted extrahepatic gene delivery can therefore benefit from surface modifications.

The objective of this work is to investigate the transfection efficacy of a library of LNPs at varying lipid molar compositions, layered with surface polyanions using layer-by-layer electrostatic deposition [3]. Negative surface polyelectrolytes can not only minimize off-target uptake, but also facilitate covalent attachment of moieties to improve endocytosis and intracellular trafficking [4]. Rationally chosen polyelectrolyte chemistries can also impact nanoparticle biodistribution when delivered intravenously, which is advantageous for extrahepatic delivery.

Circulating LNPs are currently stabilized in physiological conditions by the presence of surface polyethylene-glycol (PEG) conjugated lipids. However, PEGylated nanocarriers may induce immune responses and reduce the efficacy of subsequent nanocarrier dosages [5]. Therefore, electrostatically adsorbed polyelectrolytes that confer stability advantages while minimizing immunogenicity may provide a viable alternative to PEGylation.

Methods

We encapsulated either GFP-expressing mRNA or plasmid DNA in LNPs, then deposited carboxylated surface polyanions composed of either saccharides (hyaluronic acid, HA), polypeptides (poly-L-aspartate, PLD; poly-L-glutamate, PLE) or acrylic groups (polyacrylic acid, PAA). Transfection efficacy of unlayered and layered LNPs was assayed *in vitro* in EL4 lymphoma cells, RAW 264.7 macrophages, and HEK293T epithelial cells via flow cytometry. Layered LNP intracellular trafficking patterns were assayed via confocal microscopy studies. LNP biodistribution, transfection, and circulation times were further assessed *in vivo* in naïve C57BL/6 mice.

Results and Discussion

Relative to unlayered cationic LNPs, LNPs layered with polyanions exhibited diameter increases of 5 to 10 nanometers and complete surface charge reversal (Fig. 1A-D, F). Encapsulation efficiency of both mRNA and pDNA remained consistent between unlayered and layered LNPs (Fig. 1E). *In vitro* studies in a diverse range of immune and epithelial cell lines demonstrated correlations between lipid molar compositions and LNP transfection. Layered LNPs retained transfection capacity, transfecting at comparable or higher rates to unlayered LNPs. Furthermore, the outer polyanion of layered LNPs modulated uptake and transfection in a cell-dependent manner; HA, for example, upregulated transfection in RAW macrophages but downregulated transfection in EL4 lymphoma cells. Tested *in vivo*, PLE and HA layers further improved LNP hepatic and splenic transfection; HA-layered LNPs demonstrated splenic specificity (Fig. 2A-B). Surface deposition of polyanions further stabilized non-PEGylated LNPs in salt and serum conditions, while preserving transfection *in vivo*.

These findings suggest that LNP surface chemistry can be modulated to improve targeting and uptake, to aid in the transfection of extrahepatic tissues.

Acknowledgements: Funding was provided by the Bill and Melinda Gates Foundation and the National Science Foundation.

References

[1] Kumar et al. *Chemical Reviews* 121, 11527-11652 (2021).
[2] Samaridou et al. *Advanced Drug Delivery Reviews* 154-155, 37-63 (2020).
[3] Nabar et al. *Proceedings of the National Academy of Sciences* 121(11), (2024).
[4] Correa et al. *ACS Nano* 14(2), 2224-2237 (2020).
[5] Ju et al. *ACS Nano* 16(8), 11769-11780 (2022).

Fig. 1

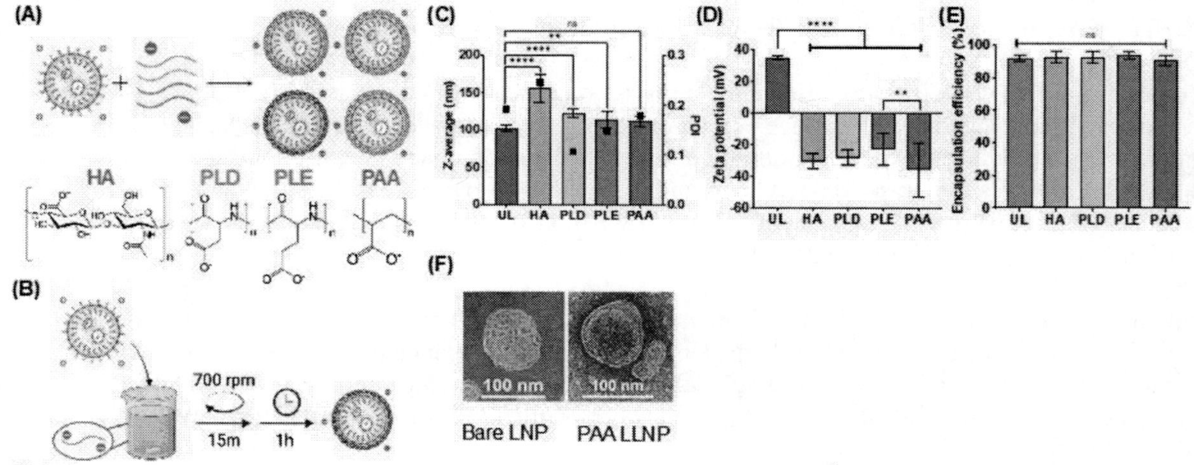

Fig. 2

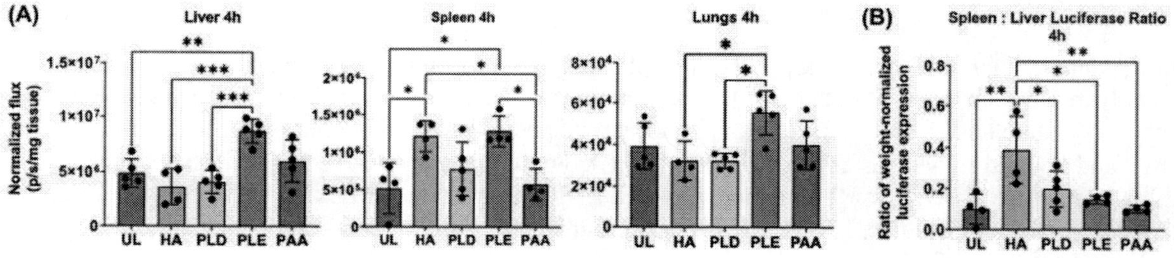

414f Two Nitrogen-Based Ionizable Lipid System for mRNA Delivery to T Lymphocytes

Joo-Youp Lee, Vishnu Sriram, Vidya Phatangare, Hyeri Sim

Delivery systems for mRNA-based therapeutics have gained enormous attention since the FDA approval of lipid nanoparticle (LNP)-based vaccines for SARS-CoV-2. LNPs are the most clinically advanced platform for nucleic acid delivery and can be formulated for various applications such as gene editing, protein replacement, and vaccine applications. Despite the enormous potential of these delivery systems, questions remain about the repeated dosage of LNPs and their potential side effects.

To address these concerns, we designed a two-nitrogen library of lipids. It was hypothesized that, at the same lipid-tertiary-amine to mRNA-phosphate (N/P) ratio, the two nitrogen atoms in the lipid would be utilized to encapsulate mRNA at a relatively low molecular weight, which in turn could reduce the total lipid dosage compared to a one-nitrogen lipid. This class of lipids could potentially mitigate the side effects posed by single-nitrogen lipid while maintaining the same performance level. We screened the library of lipids against hard-to-transfect Jurkat human T cells in vitro. We found that three lead LNPs, distinct in structure, performed well. The LNPs were optimized by evaluating the effect of cholesterol and its analogues. Furthermore, the cell uptake and endolytic pathways were determined for the lead LNPs. The three lead LNPs were then tested in vivo using Balb/c mice. Lipid 1 LNPs showed translation efficiency in the liver while Lipid 2 LNPs showed higher protein translation in spleen relative to SM-102-based LNP. Overall, the library of lipids showed promising results for mRNA delivery with reduced total lipid dosage.

416a Characterization and ROS Generation Study of Polyacrylic Acid-Coated Superparamagnetic Iron Oxide Nanoparticles for Spacecraft Water System Biofilm Control

Omid Sedighi, Eshita Jhahan, Eli Borrin, Appala Raju Badireddy, Amber Doiron

Introduction: A primary challenge in space missions is providing vital resources like water for crew members without access to Earth resupply. Water Recovery Management Systems (WRMSs) are essential to sustain life on these travels by recovering nearly 98% of the initial water source [1]. However, due to their high surface area and the moist environment, WRMSs are susceptible to biofilm formation, potentially leading to crew infection and system malfunction. Biocides are being used in many space missions to inhibit bacterial growth [2], but biocide penetration into the biofilm structure is limited by the extracellular polymeric substance (EPS). High biocide dosages are therefore required for biofilm disruption, which can result in water toxicity. This project is focused on synthesizing water-soluble, superparamagnetic iron oxide nanoparticles (SPIONPs) designed to disrupt both established and newly forming biofilms in water recovery systems, reducing the need for high doses of biocides. The anti-biofilm action of SPIONPs is multifaceted; the particles are capable of catalytic production of reactive oxygen species (ROS)—recognized for their antibacterial effects—as well as the disruption of EPS via local hyperthermia and internal shear stresses induced by oscillatory particle motion, thereby enhancing biocide penetration. A challenge with iron oxide nanoparticles is their tendency to aggregate, which negatively impacts their magnetic characteristics. To address this, polymer coatings are used to prevent aggregation. Polyacrylic acid (PAA) was chosen because of its antibacterial and biocompatible properties. Currently, we have successfully synthesized PAA-coated SPIONPs (PAA@SPIONPs) and investigated the SPIONPs' role in modulating ROS production in presence of biocides. Presence of coating on the surface of SPIONPs can potentially enhance the stability and efficacy of ROS generation over time, especially at higher temperatures.

Methods: Water-soluble, stable, and magnetically responsive SPIONPs were synthesized with the co-precipitation method which was optimized in our previously published data [3]. For the coated nanoparticles, PAA (25% w/w with respect to the iron (II) salt) was added to the reaction just before addition of ammonium hydroxide. Nanoparticle tracking analysis (NTA) and transmission electron microscopy (TEM) were employed to assess the hydrodynamic size and dry size of nanoparticles, respectively. Fourier transform infrared spectroscopy (FTIR) characterized the coating and its binding to the core SPIONPs. X-ray Photoelectron Spectroscopy (XPS) analyzed the oxidation state of the nanoparticles, crucial for determining their magnetic properties. The catalytic ability of coated and uncoated nanoparticles to generate ROS at various temperatures (21, 50, 60, 70 °C) was investigated by studying methylene blue degradation in the presence of two types of biocides: peracetic acid (PerAA), and hydrogen peroxide (H_2O_2). In this process, two types of biocides: 0.158 M PerAA and 32% w/w H_2O_2 were employed in combination with 0.6 ml of 0.5mM methylene blue and 0.464 ml of nanoparticles (1 mg/ml) for the PerAA biocide application. For the H_2O_2 biocide application, 2.32 ml of nanoparticles with the same concentration was used. The procedure was initiated by heating each test solution to a set temperature in a water bath. Upon reaching the desired temperature, all components were mixed to evaluate the reactive oxygen species (ROS) generation by the nanoparticles in the presence of each biocide. The degradation kinetics of methylene blue was evaluated for a period of 60 min. Prior to analysis of methylene blue degradation, the reaction in collected samples was halted by adding 7 ml of 1.58M sodium thiosulfate.

Results and Discussion: Our findings suggest that the addition of PAA to the synthesis step reduces the size of SPIONPs, with mean hydrodynamic sizes being 104.9±13.74 nm for uncoated SPIONPs and 84.88±3.52 nm for PAA-coated ones. Dry sizes, as measured via TEM, were 16.53±3.98 nm and 8.9±3.0 nm, respectively (Figure 1-(a,b)). TEM confirmed that our nanoparticles are within the size range (<30 nm) that is necessary for superparamagnetic behavior, which is crucial for hyperthermia applications [4], [5]. The XRD pattern (Figure 1-c) showed the six strong peaks that are correspond to crystalline planes of magnetite iron oxide nanoparticles, indicating the successful synthesis [6]. Figure 1-d shows the possible interactions between coating and iron oxide. FTIR analysis demonstrated a significant decrease in the intensity of the peak related to the COOH group on PAA, suggesting that the bonding between the coating and iron is of either type II or III. In XPS spectrum, the characteristic peaks of Fe 2p3/2 and 2p1/2 in Fe_3O_4 were observed, with the absence of a satellite line at 719 eV, indicating pure Fe_3O_4 without -Fe_2O_3 [7,8]. The temperature range of 21-70 °C is used to evaluate the effect of temperature on modulation of ROS generation by SPIONs in the presence and absence of biocides. The most effective degradation occurred at 60 °C in the presence of both biocides. Figure 1-e shows the methylene blue degradation rate using both coated and uncoated nanoparticles at 60 °C in presence of PerAA. For both coated and uncoated nanoparticles, the degradation rate was higher in the first 10 minutes, followed a decrease. The decreased degradation rate in the coated sample was less significant, resulting in a stable and prolonged degradation over time, whereas the uncoated samples exhibited almost no noticeable degradation after 10 minutes. Figure 1-f presents a similar study, but with H_2O_2 as the biocide. In this scenario, the degradation of methylene blue was more consistent for both nanoparticle types. However, the coated nanoparticles demonstrated significantly higher methylene blue degradation activity. Given our results, the presence of a coating on nanoparticles controls the release of ROS from biocides, leading to a more gradual and sustained generation of ROS, and higher methylene blue degradation rates. This indicates that the coating significantly enhances the particles' catalytic activity. The overall results demonstrated that the presence of both SPIONPs and PAA@SPIONPs could augment ROS generation in presence of biocides over time. As anticipated, the combination of SPIONPs with biocides at elevated temperature enhances their effectiveness through catalytic behavior of nanoparticles, leading to increased ROS production. Our preliminary hyperthermia studies using alternate magnetic field indicate that the temperature effects can be created to modulate ROS generation using biocides. However, we are still in the initial stages of this study, and further assessment is required before we can expose the nanoparticles to biofilms under an alternating magnetic field to study hyperthermia effects.

Conclusion: The synthesized iron oxide nanoparticles are appropriately sized and oxidized for superparamagnetic behavior. The coating has resulted in smaller, more stable nanoparticles and sustainable ROS generation at 60 °C, proving its potential for future hyperthermia studies. Future research will explore the impact of a magnetic field on ROS generation and expose these nanoparticles to bacterial cells to study the combined effects of hyperthermia and ROS generation on bacterial cells.

Acknowledgments: This work was supported by the National Aeronautics and Space Administration (NASA) under Grant No. 80NSSC21M0325.

References:

[1] P. Bernardo, A. Iulianelli, F. Macedonio, and E. Drioli, "Membrane technologies for space engineering," *J Memb Sci*, vol. 626, p. 119177, 2021.

[2] S. Paton, G. Moore, L. Campagnolo, and T. Pottage, "Antimicrobial surfaces for use on inhabited space craft: A review," *Life Sci Space Res (Amst)*, vol. 26, pp. 125–131, 2020.

[3] C. Nwasike, E. Yoo, E. Purr, and A. L. Doiron, "Activatable superparamagnetic iron oxide nanoparticles scavenge reactive oxygen species in macrophages and endothelial cells," *RSC Adv*, vol. 10, no. 68, pp. 41305–41314, 2020.

[4] H. Wei, Y. Hu, J. Wang, X. Gao, X. Qian, and M. Tang, "Superparamagnetic iron oxide nanoparticles: Cytotoxicity, metabolism, and cellular behavior in biomedicine applications," *Int J Nanomedicine*, pp. 6097–6113, 2021.

[5] C. Pucci, A. Degl'Innocenti, M. B. Gümüş, and G. Ciofani, "Superparamagnetic iron oxide nanoparticles for magnetic hyperthermia: recent advancements, molecular effects, and future directions in the omics era," *Biomater Sci*, vol. 10, no. 9, pp. 2103–2121, 2022.

[6] S. J. Iyengar, M. Joy, C. K. Ghosh, S. Dey, R. K. Kotnala, and S. Ghosh, "Magnetic, X-ray and

Mössbauer studies on magnetite/maghemite core–shell nanostructures fabricated through an aqueous route," *RSC Adv*, vol. 4, no. 110, pp. 64919–64929, 2014.

[7] G. Unsoy, S. Yalcin, R. Khodadust, G. Gunduz, and U. Gunduz, "Synthesis optimization and characterization of chitosan-coated iron oxide nanoparticles produced for biomedical applications," *Journal of Nanoparticle Research*, vol. 14, pp. 1–13, 2012.

[8] Q. Gao *et al.*, "The study of novel Fe3O4@ γ-Fe2O3 core/shell nanomaterials with improved properties," *J Magn Magn Mater*, vol. 321, no. 8, pp. 1052–1057, 2009.

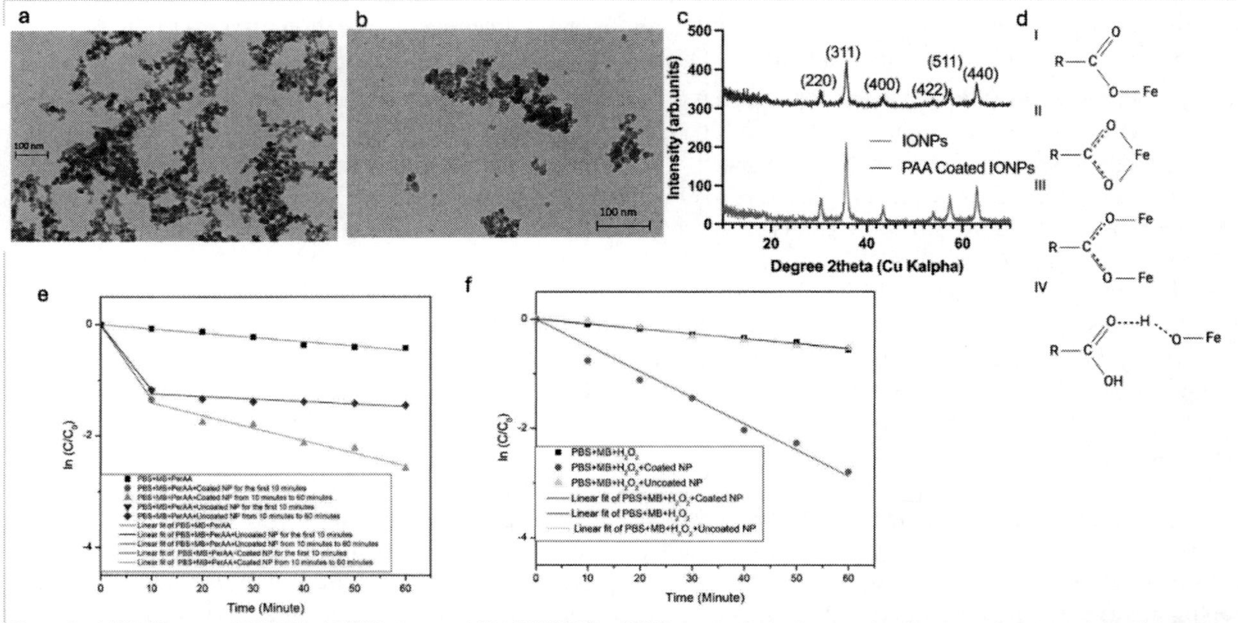

Figure 1- a) TEM Picture of SPIONPs b) TEM picture of PAA@SPIONPs c) XRD results for both coated and uncoated nanoparticles d) Possible interactions between the polymeric chains and the iron oxides e) Methylene blue degradation in presence of PerAA biocide at 60°C f) Methylene blue degradation in presence of H_2O_2 at 60°C

416b Quantifying Chirality in Multiscale Nanostructured Systems

Jessica Ma, Minjeong Cha, Ji-Young Kim, Emine Sumeyra Turali-Emre, Nicholas A. Kotov

Chirality is an essential geometric property unifying small molecules, biological macromolecules, inorganic nanomaterials, biological microparticles, and many other chemical structures. The extent of the asymmetry of chiral nanostructures influences their physical and chemical properties. Numerous chirality measures have been developed quantify this geometric property of mirror asymmetry as a way to predict or explain these properties. Many chirality measures require prohibitively demanding computations, especially for chiral structures comprised of thousands of atoms. Acknowledging the fundamental problems with quantification of mirror asymmetry, including the ambiguity of sign-variable pseudoscalar chirality measures, we revisit this subject because of their unifying significance and new aspects of chirality surfaced in nanoscale materials that display chirality continuum and scale-dependent mirror asymmetry. To address some of the fundamental problems and practical limitations of existing chirality measures, we applied the concept of torsion within the framework of differential geometry to the graph theoretical representations of molecular and nanoscale helicoids. The resulting graph-theoretical chirality measure (GTC) provides a description of both the sign and magnitude of mirror asymmetry for crystalline structures.

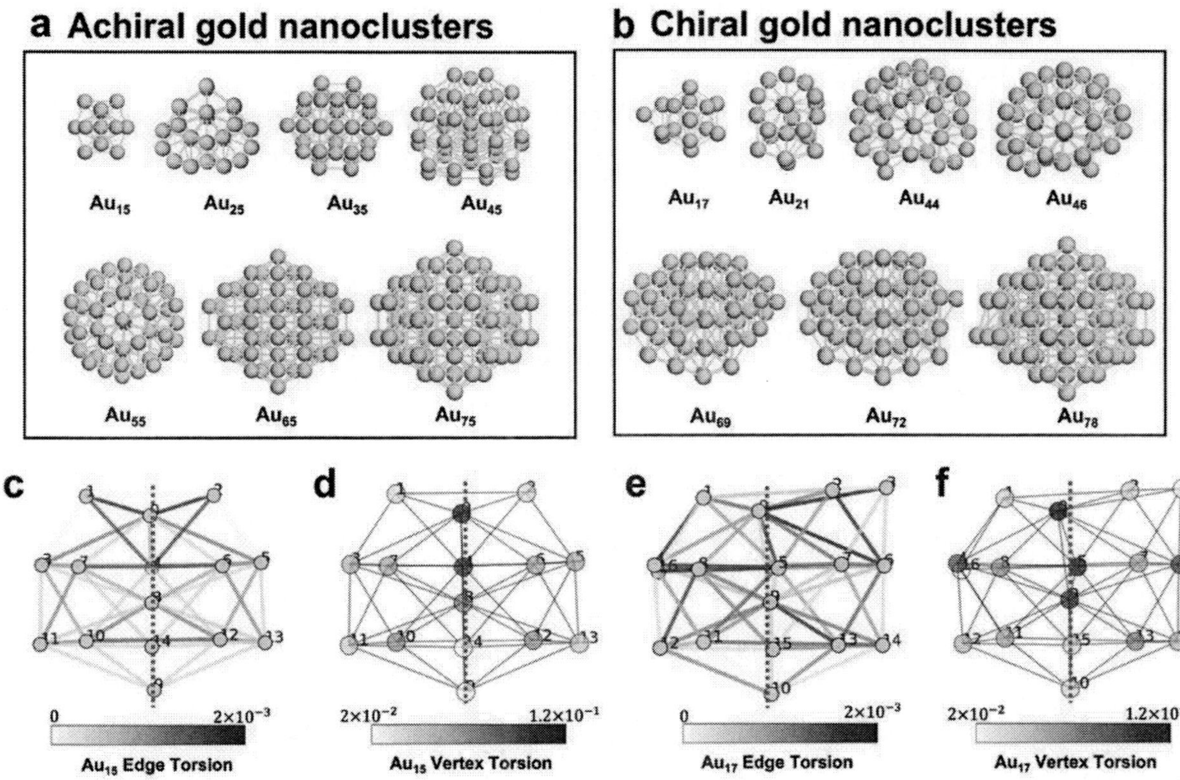

416c Role of Ligand-Induced Surface Stresses in Controlling Shapes of Ultrathin Colloidal Nanoplatelets

Sarit Dutta, Debora Monego, Doron Grossman, Asaph Widmer-Cooper, Benjamin Abécassis

Ultrathin colloidal nanoplatelets (NPL) are ligand-coated sheet-like crystalline objects with atomic scale thickness and possessing immense potential for optoelectronic applications. In solutions NPLs can adopt a variety of shapes such as cylinders, helicoids, and helical ribbons which play an important role in determining their optical properties. However the fundamental physical reasons behind the origin of multiple shapes remain poorly understood, hindering our ability to engineer NPL shapes and subsequently control their optical properties.

Here we use a combination of experiments, simulation, and theory to show that the interaction of the adsorbed ligands with the crystal surface leads to surface stresses that differ between the top and bottom surfaces and eventually forces the NPL to its ultimate deformed shape. The deformed shape is dictated by the NPL aspect ratio, thickness, orientation of the crystal with respect to the edges, and the ligand-crystal interaction, though most of the molecular details can be lumped into the spontaneous curvature. Furthermore, NPLs display a smooth transition between helicoids and helical ribbons beyond a certain critical width, a key feature of geometrically frustrated assemblies.

Ref: Monego et al (2024) Ligand-induced incompatible curvatures control ultrathin nanoplatelet polymorphism and chirality, *Proc. Natl. Acad. Sci. U.S.A.*, 121, e2316299121.

416d Engineering Alloyed 2D Transition Metal Dichalcogenide Nanosheets Via Solution Processable Methods

Rebekah Wells, Nicolas Diercks, Tian Carey, Shixin Liu, Victor Boureau, Marina Caretti, Kevin Sivula, Jonathan Coleman

The alloying of 2D TMDs is an established route to robust semiconductors with continuously-tunable optoelectronic properties. However, a major obstacle in commercializing large-area, thin devices based on alloyed TMD nanosheets is their large-scale production[1].

In this presentation we highlight our recent works addressing this issue. We begin by describing a novel powder-based, solution processable method to produce high quality 2D TMD nanosheets using a pre-annealing step and electrochemical intercalation/exfoliation[2]. Compared to traditional methods (i.e., ultrasonication), 2D TMD nanosheets produced using this method show improved optoelectronic properties thanks to high aspect ratios and low defect densities.

Next, we demonstrate how this highly adaptable method can be used to transform commercially available, pure-phase bulk TMD powders into ternary and quaternary alloyed TMD nanosheets[3]. We showcase versatility by employing this technique for two classes of alloys, $Mo_xW_{1-x}S_ySe_{2-y}$ and SnS_ySe_{2-y}. We provide evidence of the atomic mixing within the nanosheets using atomic resolution scanning transmission electron microscopy (STEM) in combination with integrated differential phase contrast (iDPC) for the metal and chalcogen atoms, respectively[4]. Furthermore, we show that control over the final composition of the nanosheets can be exerted by tuning the feed ratios of the TMD powders. Accordingly, we examine the unique electronic and optoelectronic properties that arise as a function of the chemical composition of the alloy. Notably, the phenomena observed are consistent with nanosheets produced *via* chemical vapor deposition (CVD) and related methods, suggesting that this versatile method is an economically viable solution for making alloyed 2D TMD nanosheets. Indeed, the ability to produce composition controlled TMD nanosheets in large quantities is critical for the inexpensive production of next-generation, large-area optoelectronic devices.

References:

(1) Yu, X.; Sivula, K. Toward Large-Area Solar Energy Conversion with Semiconducting 2D Transition Metal Dichalcogenides. *ACS Energy Lett.* **2016**, *1* (1), 315–322. https://doi.org/10.1021/acsenergylett.6b00114.

(2) Wells, R. A.; Zhang, M.; Chen, T.-H.; Boureau, V.; Caretti, M.; Liu, Y.; Yum, J.-H.; Johnson, H.; Kinge, S.; Radenovic, A.; Sivula, K. High Performance Semiconducting Nanosheets *via* a Scalable Powder-Based Electrochemical Exfoliation Technique. *ACS Nano* **2022**, *16* (4), 5719–5730. https://doi.org/10.1021/acsnano.1c10739.

(3) A. Wells, R.; J. Diercks, N.; Boureau, V.; Wang, Z.; Zhao, Y.; Nussbaum, S.; Esteve, M.; Caretti, M.; Johnson, H.; Kis, A.; Sivula, K. Composition-Tunable Transition Metal Dichalcogenide Nanosheets via a Scalable, Solution-Processable Method. *Nanoscale Horizons* **2024**. https://doi.org/10.1039/D3NH00477E.

(4) Lazić, I.; Bosch, E. G. T.; Lazar, S. Phase Contrast STEM for Thin Samples: Integrated Differential Phase Contrast. *Ultramicroscopy* **2016**, *160*, 265–280.

https://doi.org/10.1016/j.ultramic.2015.10.011.

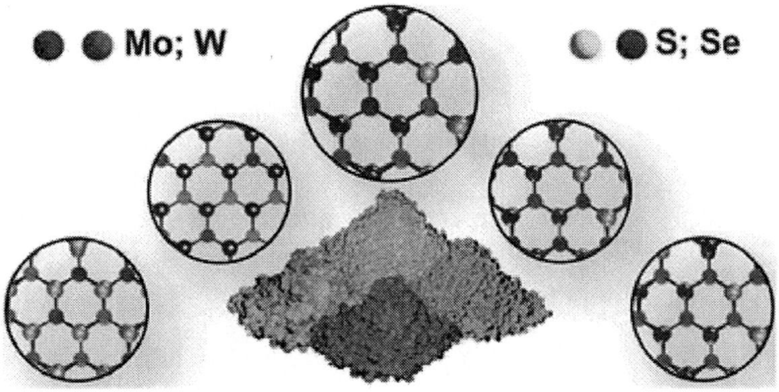

416e Electromagnetic Coupling and Transport in a Topological Insulator–Graphene Heterostructure

Enrique Munoz

The combination of different materials in the form of heterostructures [1] is our modern quest to search for novel properties that emerge beyond the trivial superposition of those of their individual parts. This search is of great interest not only from a fundamental perspective, since exciting new phenomena may be observed, but also to engineer materials for applications in novel devices. Among the different emerging phenomena in heterostructures, electromagnetic effects are highly relevant for the transmission and storage of energy and information. In this context, the control of electronic transport properties is of fundamental importance. The discovery of novel materials with nontrivial topological properties, such as topological insulators (TIs) [1,3,4] and Dirac and Weyl semimetals [2], has introduced a plethora of new phenomenology. In particular, the existence of surface chiral states makes them excellent potential candidates for applications in quantum information technologies and thermoelectrics [2,3,4]. In addition, the so-called magnetoelectric polarizability (MEP) [1] that locally modifies the constitutive relations between the electromagnetic fields in TIs provides new opportunities to control the magnetoelectric response in such systems. Even though these effects have been extensively studied in individual topological materials, their electromagnetic coupling when integrated into heterostructures remains a vast territory for further exploration.

In this work [1], we consider a heterostructure composed of a TI slab and a single graphene layer, as depicted in Fig. 1. We further assume that a diluted concentration of ionized impurities is present in the graphene monolayer. The presence of such charged impurities will induce a local distortion of the charge density of the 2D electron gas, leading to a nontrivial electromagnetic coupling between the TI and the graphene monolayer in the heterostructure. As a probe of this coupling, we further studied the electrical conductivity as a function of temperature, by including the scattering effects with the local electromagnetic field configuration via the Kubo linear response formalism [1,2]. We applied our theoretical results to model the electromagnetic coupling in heterostructures made of different TIs (PbTe, Bi2Te3, PbSe, PbS, Bi2Se3, TlBiSe2, TbPO4). Our analytical and numerical results suggest that, among the properties of the TIs, the dielectric permittivity $\varepsilon 1$ is the most relevant at tuning the electronic transport in the coupled graphene monolayer. On the other hand, we also observed that the topological effects arising from the presence of the MEP coefficient θ are comparatively small even at zero temperature.

References:

[1] D. Bonilla, J. Castaño-Yepes, A. Martín-Ruiz, and E. Muñoz, Physical Review B 107, 245103 (2023) https://doi.org/10.1103/PhysRevB.107.245103

[2] D. Bonilla and E. Muñoz, Nanomaterials 12, 3711 (2022) https://doi.org/10.3390/ nano12203711

[3] J. Castaño-Yepes and E. Muñoz, Annalen der Physik 202300398 (2024) https://doi.org/10.1002/andp.202300398

[4] J. Castaño-Yepes and E. Muñoz, Results in Physics 39, 105712 (2022) https://doi.org/10.1016/j.rinp.2022.105712

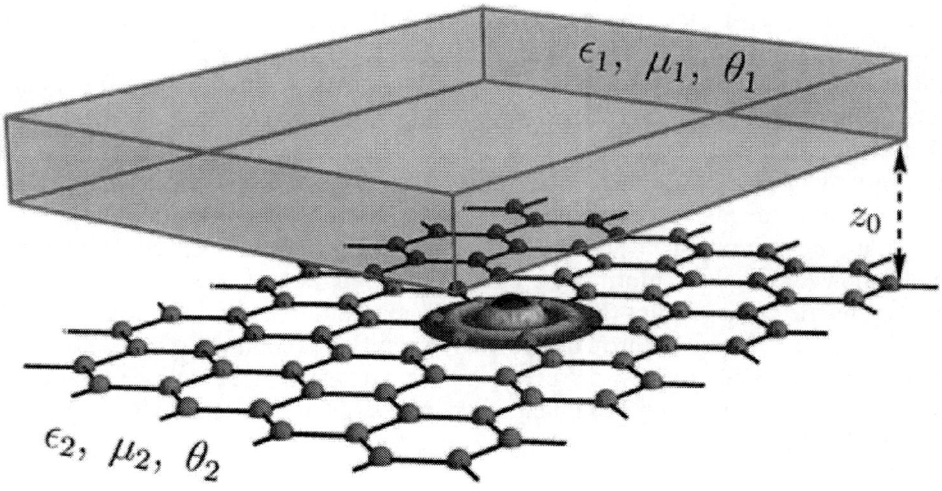
$\epsilon_1,\ \mu_1,\ \theta_1$
z_0
$\epsilon_2,\ \mu_2,\ \theta_2$

416f Linker Engineering for DNA Superstructure Design

Tobias Dwyer, Zohar Arnon, Dan Redeker, Daniel McKeen, Dayoung Lee, Timothy C. Moore, Sharon C. Glotzer, Oleg Gang

DNA is a molecule with a large engineering potential due to its well understood, designable interactions. In the past decade, nanoscale building blocks assembled entirely out of DNA have been synthesized through 'DNA-origami' methods. These building blocks have simple, complementary interactions that can be engineered to assemble a variety of superlattices. However, the study of DNA origami systems has been mostly limited to simple single origami building block assemblies. To unlock the full potential of the DNA design space to engineer complex superstructures, we seek to combine multiple DNA origami building block shapes and predict their co-assembly. **In this work,** we model the self-assembly of DNA origami "nanocages" (wireframe polyhedra) self-assembled with complementary double-stranded DNA linkers. The DNA origami nanocages are not complementary to each other but are instead linked together with double-stranded DNA containing sticky ends complementary to DNA strands on the vertices of the nanocages. Using coarse grained molecular dynamics simulations implemented with HOOMD-Blue, we show how small changes in the DNA linker – origami interactions can lead to surprising changes in the crystal structures observed in simulation and experiment. Our simulations elucidate that differences in nanoscale binding preferences drive the observed differences in crystal structure and allow us to engineer crystal structures for multiple linker-mediated DNA origami nanocage assemblies. This combined simulation, experimental work, thereby paves the way for future efforts in engineering final DNA superstructures and DNA lattices that can reconfigure via in-situ modification of nanoscale interactions.

416g Intershell Locking of Double-Walled Carbon Nanotubes Resulting from Enhanced Intershell Friction at High Shear Rates

Hongjie Yue, Fei Wei

Friction is a pervasive phenomenon in various mechanical processes, especially in systems involving high-speed motion where friction has a great impact on energy dissipation and mechanical transmissions. Meanwhile, friction in some nanomaterials exhibits velocity-dependent phenomena that are different from macroscopic friction, suggesting that friction has a significant impact on nanoscale systems involving highspeed motion. However, the measurement of nanoscale-friction at high sliding velocities has remained elusive due to the limitations in temporal and spatial resolution associated with nanomanipulation techniques in electron microscopes. Here, we developed a non-contact photocatalytic cut-off method specifically tailored for the outer shells of double-walled carbon nanotubes (DWCNTs), and captured the motion of the outer shells as they slide along the inner shells subsequent to fracture under an optical microscope. The result indicates that the intershell friction is independent of the overlap length but linearly related to the sliding velocity. At velocity of 977 mm/s or shear rate of 109 s-1, the intershell friction reaches 195 nN——comparable in scale to tensile stress. Moreover, the linear increase of shear stress with shear rate presents a quasi-Newtonian fluids sliding behavior, wherein the shear viscosity exhibits a decrease from 2 mPa·s to 0.06 mPa·s and then maintains a relatively constant value at high shear rates. Such enhanced intershell friction leads to locking of the inner and outer tubes when the DWCNTs undergo high-speed fracture, thus improving the resistance of DWCNTs to high-speed impacts.

212a Nanoporous 3D Graphene-like Carbons for Electrochemical Energy Storage

Hongjun Park, Raj Kumar Bera, Ryong Ryoo

In this presentation, we showcase recent advancements in energy storage applications utilizing zeolite-templated carbon (ZTC) materials [1]. The ZTC synthesis involves metal-ion-catalyzed carbonization followed by the removal of zeolite templates [2]. Successful synthesis yields ZTC products characterized by well-ordered microporous structures boasting high surface areas (> 2500 m^2g^{-1}) and remarkably narrow pore size distributions, typically around 1 nm [1]. This unique structure finds interesting application in various energy storage systems, including supercapacitors, lithium-ion batteries and capacitors [2–5]. Moreover, the nanopores of ZTC can accommodate guest species such as Co nanoparticles, molecular I_2, and MnO_x nanostructures, encapsulating them for prolonged cycling stability [6–8]. As reported in the literature, ZTC serves as an intriguing electrode material extending beyond traditional gas storage applications.

References

1. H. Nishihara and T. Kyotani, *Chem. Commun.,* **2018**, *54*, 5648

2. K. Kim *et al., Nature,* **2016**, *404*, 760–770

3. Y. Kwon *et al., J. Phys. Chem. C,* **2018**, *122*, 4955

4. H. Park *et al, Carbon,* **2019**, *155*, 570–579

5. H. Park *et al., submitted*

6. R. K. Bera *et al., ChemElectroChem,* **2021**, *8*, 4038

7. H. Park *et al., Adv. Energy Sustainability Res.,* **2021**, *2*, 2100076

8. S. Khan *et al., Micropor. Mesopor. Mater.,* **2024**, *369*, 113050

212b Tailored Growth and Manipulation of Atomically Thin Layered Materials for Optoelectronics, Energy Harvesting and Beyond

Debjit Ghoshal

An increased thrust for smarter and energy efficient devices has led to the search for newer materials with enhanced functionalities. Graphene has taken center stage in the lively area of electronics. Monolayers of 2D transition metal dichalcogenides due to their tunable properties, have shown immense promise in applications such as optoelectronics, flexible electronics and spintronics. However, despite the enormous potential shown by atomically thin sheets of these layered 2D materials, they are far from commercialization. A major roadblock has been the lack of scalable techniques for synthesis and manipulation of these materials. In this presentation, the focus will be on some unique and upcoming approaches to grow and defect engineer wafer scale epitaxial 2D materials which can drastically improve their functionalities. Other approaches to potentially tune the properties of these materials like doping and phase engineering are discussed. We envision the ability to scalable fabrication and modulation these materials in a controllable fashion can open-up additional knobs to tune their properties, thus making them more marketable for various applications. Further, future directions, funding possibilities and possible collaboration opportunities going forward for applications of these defect engineered/phase modified 2D materials in various upcoming applications like membrane separation, as single photon emission (SPE) sources, remote epitaxy and energy harvesting will be highlighted and discussed.

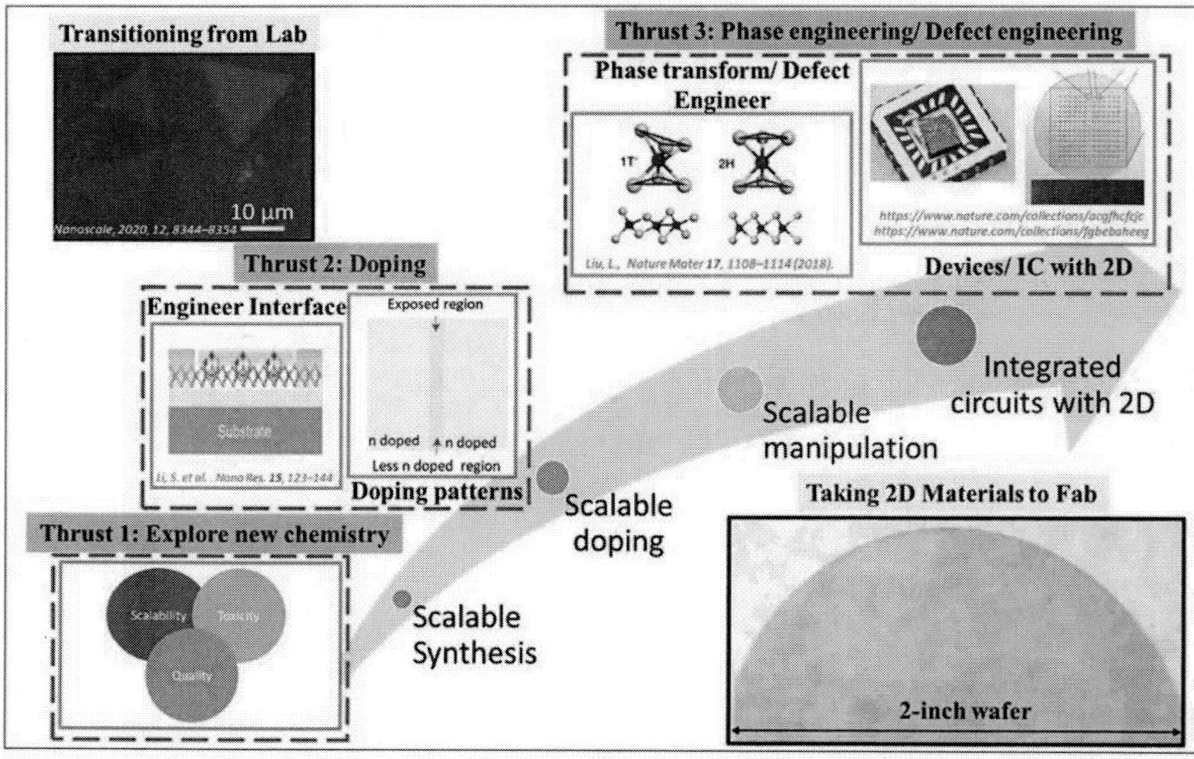

212c Heterogeneity, Dynamic Ordering, and Ion Transport: Emerging Methods for Characterizing Complex Solids at the Nanoscale

Julian Vigil

Complex structure—characteristic of engineered nano- and micro-scale materials—often presents challenges to rigorous characterization using conventional methods, including microscopy and diffractometry. Here, emerging measurement and analysis methods are highlighted, primarily toward rationalizing the properties of functional materials for optoelectronic devices and renewable-energy applications. The advent of next-generation light sources and detector technologies have enabled new capabilities such as rapid *or* low-dose acquisition, pump-probe ultrafast modalities, and enhanced spatial resolution.

Using metal-halide perovskite semiconductors as case studies, we demonstrate how applied crystallography methods elucidate: (*i*) compositional heterogeneity and anomalous phase behavior in perovskite polycrystals (solution-processed thin films); (*ii*) intrinsically dynamic local order within the average high-symmetry structures of perovskite single crystals. The suite of synchrotron X-ray and neutron scattering methods serve to differentiate the static and dynamic disorder within these complex hybrid organic-inorganic crystals. Connections between measurement and calculations—particularly contrasting equilibrium *elastic* and energy-filtered *inelastic* scattering—will also be discussed toward atomic-scale modeling of complex materials.

Finally, scanning transmission electron microscopy (STEM) and electron nanoprobe diffraction (e.g., 4D-STEM) techniques are developed to characterize dynamic structure-transport relations. To investigate atomic-scale instabilities in the metal-halide perovskites, we used cryogenic 4D-STEM to directly map phase segregation due to photoinduced halide mobility. Further, room-temperature and high-temperature *in situ* 4D-STEM measurements elucidate transport mechanisms in graphite intercalation compounds for applications from battery electrodes to 2D electronics. With resolution limits realistically approaching the atomic scale, STEM-based methods to differentiate reversible and irreversible transformations, as well as diffusion-reaction boundaries, are proposed.

212d Understanding the Electrodeposition Mechanism of Chiral Cu Catalyst for the Efficient Electrochemical Carbon Dioxide Reduction

Jeiwan Tan, Nathan Neale, Jao van de Lagemaat

In the past years, the concept of chirality-induced spin selectivity (CISS) has been shown to improve the efficiency of oxygen evolution reactions by stabilizing only one spin state of charge carriers at the catalyst surface leading to a reduction in the efficiency of peroxide and singlet oxygen formation vs triplet oxygen formation. However, the detailed role of chiral organic molecules during the electrodeposition of metal/metal oxide electrocatalysts have not been investigated. In this work, we develop *in situ* spec-echem techniques to understand the in-depth mechanism of the enantiospecific electrodeposition and demonstrate how the CISS phenomenon can be used to control the reactivity and selectivity at the cathodic side by employing chiral Cu-based electrocatalysts. The onset potential of HER on chiral Cu catalysts negatively shifts compared to the achiral counterpart despite very similar physical and chemical properties of catalyst films. The electrochemical impedance spectroscopy suggests an altered reaction pathway and more sluggish HER kinetics on chiral Cu catalysts, leading to high Faradaic efficiency and selectivity of desired carbon products. Our findings provide insight into the potential of chiral catalysts for controlling the selectivity during CO2R and other valuable reduction reactions.

212e Designing Charged Membranes for Energy Applications

Joan Montes de Oca

The increasing demand for renewable energy sources and improved energy efficiency underscores the critical need for innovative materials and technologies to address these concerns. Membranes with advanced functionalities stand out for their potential to revolutionize energy storage in flow batteries, enable osmotic power extraction, improve water desalination processes, and more. This presentation, part of the faculty candidate series, will highlight my contributions to developing nanoscale materials for energy applications during my tenure as a postdoctoral scholar at the University of Chicago and as a member of the AMEWS Energy Frontier Research Center.

I will begin by discussing design strategies for mesoporous membranes for osmotic power extraction. I will examine the molecular mechanisms that limit osmotic power harvesting and some strategies we envision could address these obstacles. In particular, I will focus on the importance of ionic current rectification (ICR), its origins, and methodologies for enhancing ICR without sacrificing total power output.

Subsequently, I will review my contributions to understanding the relationship between water uptake and ionic conductivity in ion-exchange membranes (IEMs). I will address what we found to be a near-universal correlation between water's hydrogen bond network percolation and the emergence of continuous low electric resistance pathways. Finally, I will propose possible approaches to decouple hydration from conductivity, potentially unlocking new avenues for enhancing IEM performance.

To conclude, I will briefly outline my scientific vision for the future of membrane technology in energy-water applications, emphasizing potential research directions and innovations I will explore as a new faculty member.

212f Bright and Stable Light-Emitting Diodes Made with Perovskite Nanocrystals Stabilized in Metal–Organic Frameworks

Hsinhan Tsai, Wanyi Nie

Perovskite nanocrystals are exceptional candidates for light-emitting diodes (LEDs). However, they are unstable in the solid film and tend to degrade back to the bulk phase, which undermines their potential for LEDs. Here we demonstrate that perovskite nanocrystals stabilized in metal–organic framework (MOF) thin films make bright and stable LEDs. The perovskite nanocrystals in MOF thin films can maintain the photoluminescence and electroluminescence against continuous ultraviolet irradiation, heat and electrical stress. As revealed by optical and X-ray spectroscopy, the strong emission originates from localized carrier recombination. Bright LEDs made from perovskite-MOF nanocrystals are demonstrated with a maximum external quantum efficiency of over 15% and a high brightness of over 10^5 cd m^{-2} after the device reaches stabilization. During LED operation, the nanocrystals can be well preserved, free of ion migration or crystal merging through protection by the MOF matrix, leading to a stable performance over 50 hours.

AUTHOR INDEX

AIChE
120 Wall Street, FL 23
New York, NY 10005-4020

ISBN 979-8-3313-1681-5
9 798331 316815

CURRAN ASSOCIATES INC.
proceedings.com